NEW APPROACHES TO CIRCLE PACKING IN A SQUARE

With Program Codes

Optimization and Its Applications

VOLUME 6

Aims and Scope

Optimization has been expanding in all directions at an astonishing rate during the last few decades. New algorithmic and theoretical techniques have been developed, the diffusion into other disciplines has proceeded at a rapid pace, and our knowledge of all aspects of the field has grown even more profound. At the same time, one of the most striking trends in optimization is the constantly increasing emphasis on the interdisciplinary nature of the field. Optimization has been a basic tool in all areas of applied mathematics, engineering, medicine, economics and other sciences.

The series *Optimization and Its Applications* publishes undergraduate and graduate textbooks, monographs and state-of-the-art expository works that focus on algorithms for solving optimization problems and also study applications involving such problems. Some of the topics covered include nonlinear optimization (convex and nonconvex), network flow problems, stochastic optimization, optimal control, discrete optimization, multi-objective programming, description of software packages, approximation techniques and heuristic approaches.

NEW APPROACHES TO CIRCLE PACKING IN A SQUARE

With Program Codes

By

P.G. SZABÓ
University of Szeged, Hungary

M.Cs. MARKÓT
Hungarian Academy of Sciences, Budapest, Hungary

T. CSENDES
University of Szeged, Hungary

E. SPECHT
Otto von Guericke University of Magdeburg, Germany

L.G. CASADO
University of Almería, Spain

I. GARCÍA
University of Almería, Spain

Springer

Additional material to this book can be downloaded from http://extras.springer.com.

ISBN-10: 1-4899-8897-1 ISBN 0-387-45676-7 (eBook)

ISBN-13: 978-1-4899-8897-3

Printed on acid-free paper.

Softcover re-print of the Hardcover 1st edition 2007

9 8 7 6 5 4 3 2 1

springer.com

Contents

Preface

The problem of finding the densest packing of congruent circles in a square is obviously an interesting challenge of discrete and computational geometry with all its surprising structural forms and regularities. It is easy to understand what the problem is: to position a given number of equal circles in such a way that the circles fit fully in a square without overlapping.

With a large number of circles to be packed, the solution is very difficult to find. This difficulty is highlighted by many features. On one hand it is clear that an optimal solution can be rotated, reflected, or the circles can be reordered, and hence the number of equivalent optimal solutions blows up as the number of circles increases. On the other hand, in several cases there even exists a circle in an optimal packing that can be moved slightly while retaining the optimality. Such free circles (or rattles) mean that there exist not only a continuum of optimal solutions, but the measure of the set of optimal solutions is positive!

One nice aspect of circle packing is that the problem is clear without any further explanation, and a solution can be provided by a figure. In a certain sense, the number of circles for which we know the optimal packing with certainty indicates how sophisticated the available theoretical and computational tools are. One can think that it is a kind of measure of our technological and scientific capabilities.

Beyond the theoretically challenging character of the problem, there are several ways in which the solution methods can be applied to practical situations. Direct applications are related to cutting out congruent two-dimensional objects from an expensive material, or locating points within a square in such a way that the shortest distance between them is maximal. Circle packing problems are closely related to the 'obnoxious facility location' problems, to the Tammes problem (locate a given number of points on a sphere in such a way that the minimal distance among them is maximal), and less closely related to the Kissing Number Problem (determine how many unit spheres can touch a given unit sphere in an n-dimensional space). The emerging computational algorithms can also be well utilized in other hard-to-solve optimization problems like molecule conformation.

The involvement of the authors with circle packing began in 1993 when one of the authors, Tibor Csendes, visited the University of Karlsruhe in Germany. His colleague, Dietmar Ratz, showed him a paper written in the IBM Nachrichten (42(1992) 76–77) about the difficulties involved in packing ten circles in the unit square. The problem seemed to fit the interval arithmetic-based reliable optimization algorithms they were investigating.

However, the first tries were disappointing: not even the trivial three-circle problem instance could be solved with the hardware and software environments available at that time. It became clear that symmetric equivalent solutions made the problem hard to solve.

A couple of years later a PhD student, Péter Gábor Szabó, asked Tibor Csendes for an operations research subject. Among those offered was the circle packing problem, since Péter graduated as a mathematician (and not as a computer science expert, as do the majority of the PhD students at the Institute of Informatics of the University of Szeged). He began with an investigation of the structural questions of circle packing.

The Szeged team collaborated with the Department of Computer Architecture and Electronics at the University of Almería in Spain in optimization, first in the framework of a Tempus project, and then for some years within a European Erasmus / Socrates programme. The head of the Spanish department, Inmaculada García, sent two PhD students to Szeged for a few months. One of the PhD students, Leocadio González Casado, was interested in the circle packing problem and invested some time in a new approximating solution algorithm. Péter Gábor Szabó visited the Almería team, and the joint work resulted in the first double article reporting six world best packings at that time. These candidate optimal packings were all beaten later by the results of Eckard Specht at the University of Magdeburg. His collaboration with the Hungarian team brought other publications on the subject.

The last member joining the present team was Mihály Csaba Markót. He was a PhD student supervised by Tibor Csendes, but was working on another topic, improving the efficiency of interval arithmetic-based global optimization algorithms. His interest turned relatively late to circle packing, when the interval optimization methods seemed to be effective enough to tackle such tough problems. It was very interesting to see what kind of utilization of the problem specialities was necessary to build a numerically reliable computer procedure that cracked the next unsolved problem instances of 28, 29, and 30 circles. That produced a kind of happy end for this nearly 10-year story—and a return to the first hopeless-looking technique.

The authors had intended this volume to be a summary of results achieved in the past few years, providing the reader with a comprehensive view of the theoretical and computational achievements. One of the major aims was to publish all the programming codes used. The checking performed by the wider scientific community has helped in having the computational proofs accepted. The open source codes we used will enable the interested reader to improve on them and solve problem instances that still remain challenging, or to use them as a starting point for solving related application problems.

The present book can be recommended for those who are interested in discrete geometrical problems and their efficient solution techniques. The volume is also worth reading by operations research and optimization experts as a report or as a case study of how utilization of the problem structure and specialities enabled verified solutions of the previously hopeless high-dimensional nonlinear optimization problems with nonlinear constraints. The outlined history of the whole solution procedure provides a balanced picture of how theoretical results, like repeated patterns, lower and upper bounds on possible optimum values, and approximate stochastic optimization techniques, supported the final resolution of the original problem.

Acknowledgments. The authors would like to thank all their colleagues who participated in the research. This work was supported by the Grants OTKA T 016413, T 017241, T 034350, FKFP 0739/97, and by the Grants OMFB D-30/2000, OMFB E-24/2001. It was also supported by the SOCRATES-ERASMUS programme (25/ ERMOB/1998-99), by the Spanish Ministry of Education (CICYT TIC96-1125-C03-03) and by the Consejería de Educación de la Junta de Andalucía (07/FSC/MDM). All grants obtained are gratefully acknowledged, and the authors hope they have been used in an efficient and effective way – as in part reflected by the present volume. Mihály Csaba Markót would like to thank the Advanced Concepts Team of the European Space Agency, Noordwijk, The Netherlands for the possibility to prepare the manuscript during his postdoctoral fellowship. The authors are grateful to Jose Antonio Bermejo (University of Almería, Spain) for preparation of the enclosed CD-ROM, and to David P. Curley for checking this book from a linguistic point of view.

Glossary of Symbols

n	the number of circles/points to be packed/arranged, page 1
$P(r_n, S)$	a circle packing, page 13
r_n	the common radius of a circle packing, page 13
S	the size of the enclosing square of a circle packing, page 13
$\overline{r}_n$	the optimum value of the circle packing problem, page 13
$A(m_n, \Sigma)$	a point arrangement, page 14
m_n	the minimal pairwise distance in a point arrangement, page 14
Σ	the size of the enclosing square for a point arrangement, page 14
$\overline{m}_n$	the optimum value of the point arrangement problem, page 14
$P'(R, s_n)$	an associated circle packing, page 14
s_n	the size of the enclosing square for an associated circle packing, page 14
R	the common radius for an associated circle packing, page 14
$\overline{s}_n$	the optimum value for an associated circle packing, page 14
$A'(M, \sigma_n)$	an associated point arrangement, page 14
σ_n	the size of the enclosing square for an associated point arrangement, page 14
M	the minimal pairwise distance for an associated point arrangement, page 14
$\overline{\sigma}_n$	the optimum value for an associated point arrangement problem, page 14
$\mathcal{P}_i^n$	Problem 2.i ($1 \leq i \leq 5$), page 14
$d_n(r_n, S)$	the density of a circle packing $P(r_n, S)$, page 17
T_h	threshold level in Threshold Accepting Algorithm, page 33
ξ	the variance of the perturbation size, page 35
d_{ij}	the squared distance between two points for a point arrangement, page 52
$d(x, y)$	the squared distance between $x \in \mathbb{R}^n$ and $y \in \mathbb{R}^n$, page 52
$f_n(x, y)$	the objective function for Problem 6.2, page 52
$\mathbb{I}$	the set of real, compact intervals, page 53
$\text{lb}(A)$	the lower bound of the interval A, page 53
$\text{ub}(A)$	the upper bound of the interval A, page 53
$\mathbb{I}(S)$	the set of intervals in $S \subseteq \mathbb{R}$, and the set of boxes in $S \subseteq \mathbb{R}^n$, page 53
$w(A)$	the width of the interval A, page 53
$F(X)$	an interval inclusion function of a real function f over the box X, page 53

$f(X)$	the range of a real function f over the box X, page 53
D_{ij}	the natural interval extension of d_{ij}, page 56
$F_n(X,Y)$	an interval inclusion function of $f_n(x,y)$, page 56
$\tilde{f}$	the best known lower bound for the maximum of Problem 6.2, page 57
$\tilde{f}_0$	the best known lower bound for the maximum of Problem 6.1, page 57
$D(x,y)$	the natural interval extension of $d(x,y)$, page 79
$S^m_{s..f}$	input sets of tile combinations for the B&B method, page 89
$\bar{S}^m_{s..f}$	output sets of tile combinations for the B&B method, page 89
${}^{m_1}_{s_1..f_1}S^m_{s..f}$	input sets of tile combinations for the B&B method, page 89
${}^{m_1}_{s_1..f_1}\bar{S}^m_{s..f}$	output sets of tile combinations for the B&B method, page 89
PAT$(f(k))$	a pattern class, where $n = f(k)$, page 115
STR$(f(k))$	a structure class, where $n = f(k)$, page 123
$[[p,q]]$	a grid packing, page 132
GP	the set of grid packings, page 132
$p^I_n(x)$	a generalized minimal polynomial, page 145
$P_n(x)$	$p^1_n(x)$, page 145

1

Introduction and Problem History

1.1 Problem description and motivation

The general mathematical problem of finding the densest packing of equal objects in a bounded shape arises in many fields of the natural sciences, in engineering design, and also in everyday life. Some applications that involve the packing of *identical* circles are the following:

- *coverage*: place radio towers in a geographical region such that the coverage of the towers is maximal, with as little interference as possible;
- *storage*: place as many identical objects as possible (e.g. barrels) into a storage container;
- *packaging*: determine the smallest box into which one can pack a given number of bottles;
- *tree exploitation*: plant trees in a given region such that the forest is as dense as possible, but the trees allow each other to grow up to their maximal desired size;
- *cutting industry*: cut out as many identical disks as possible from a given (in the general case, irregular) piece of material.

All the above applications require the solution of the following general problem:

Place $n \geq 2$ identical circles (disks) in a given, bounded subset of the plane without overlapping, in such a way that the density of the packing is maximal. (The density of the packing is defined as the ratio of the filled and total available area.)

In this book we mainly deal with the problems of packing equal circles in the square. However, some of the results (especially, the stochastic algorithms of Chapters 4 and 5) can be applied or generalized to more general packing scenarios as well.

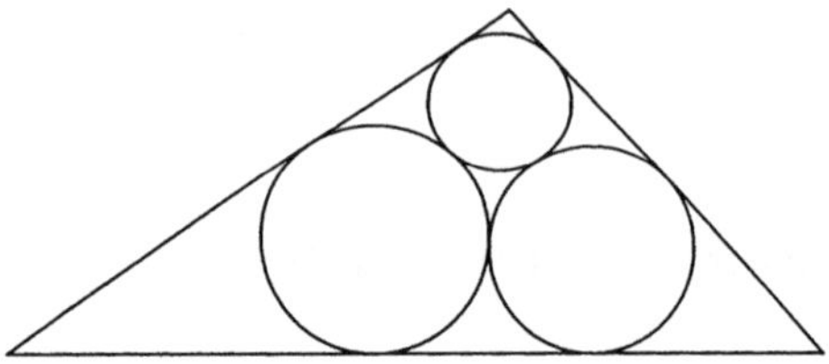

Fig. 1.1. The Malfatti circles.

1.2 The history of circle packing

The circle packing problem has a long history in mathematics, see e.g. [94, 118, 130]. In this section we will review some of the existing results, starting from the early investigations.

1.2.1 The problem of Malfatti

In European mathematics, one of the first examples of looking for some kind of optimal packing of circles is the famous *'Malfatti Problem'*. In 1803, the Italian mathematician Gianfrancesco Malfatti (1731–1807) posed the following question: *Consider a right prism with a right triangular base. How do we cut out three cylinders (perhaps of different sizes) from the prism, such that the total volume of the cylinders is maximal?*

Malfatti thought that the solution was to construct three circles in the triangle, in such a way that each one of them touched the other two circles and two sides of the triangle. In mathematical literature, the geometric construction of such circles in an *arbitrary* triangle is called the Malfatti Problem (Figure 1.1). However, it is worth mentioning that this latter problem was previously studied and solved earlier by the Japanese mathematician Chokuen Ajima (1732–1798) [30].

Interestingly, it turned out only about a hundred years later (according to our knowledge) that the Malfatti circles do not necessarily give the densest packing of three (non-identical) circles in a triangle: In 1930, H. Lob and H. W. Richmond [59] showed that in an equilateral triangle a different packing can give a greater density than those of the Malfatti circles (see Figure 1.2).

The problem of finding the densest possible configuration in an arbitrary triangle was solved in 1992 by V. A. Zalgaller and G. A. Los' [131].

1.2.2 The circle packing studies of Farkas Bolyai

It seems likely that the Hungarian mathematician Farkas Bolyai (1775–1856) was the first scientist who investigated the density of circle packing *sequences* in bounded regions and published the related results. In his main work – usually referred to as the 'Tentamen', 1832–33 [6] –, a dense packing of equal circles in an equilateral triangle was worked out, as depicted in Figure 1.3.

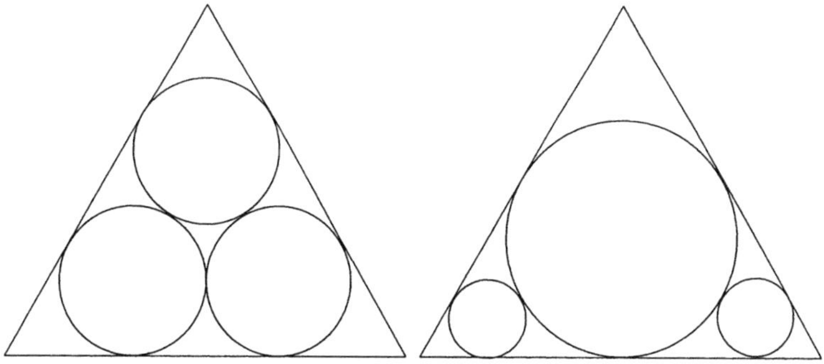

Fig. 1.2. The Malfatti circles and a denser packing in an equilateral triangle.

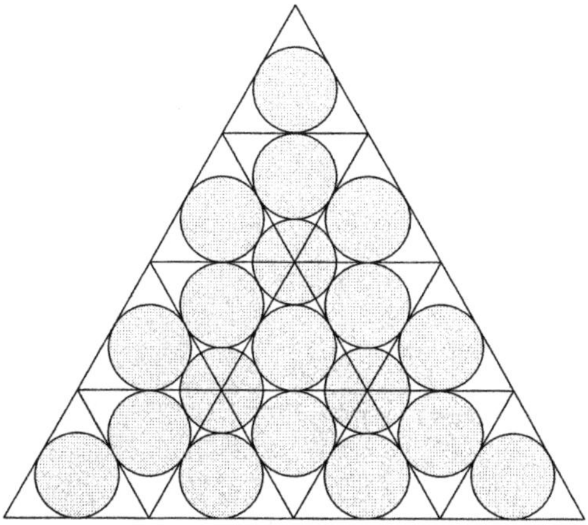

Fig. 1.3. The example of Bolyai for packing 19 equal circles in an equilateral triangle.

This particular packing is obtained by dividing the triangle into congruent equilateral triangles and placing the circles in these smaller triangles. Then three additional circles with the same radius can be placed in the positions which are surrounded by six circles.

Bolyai defined an infinite packing sequence based on the refinement of the subdivision of the large triangle, and investigated the limit of the *'vacuitas'* (in Latin, the area not covered by the circles). It is easy to demonstrate that the limit density of the resulting packing sequence is $\pi/\sqrt{12}$.

An interesting historical question is the motivation behind the studies of Bolyai in this subject. According to research of mathematical history, he was about to apply for a position in the forestry commission, and this led him to the investigation of problems like planting trees in given regions such that 'they share the same amount of light and air'. Figures 1.4 and 1.5 show some of his other packing examples in this

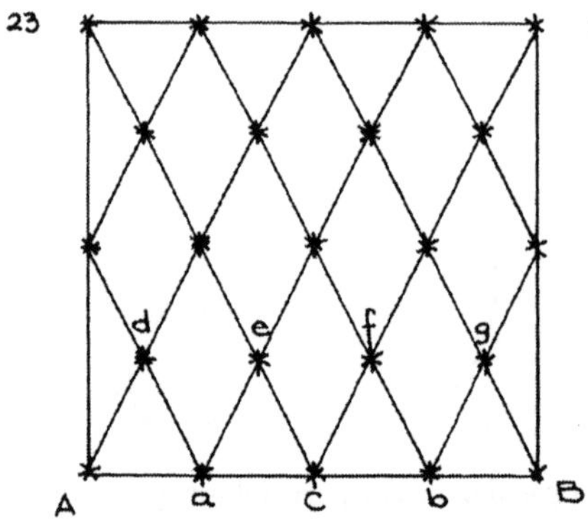

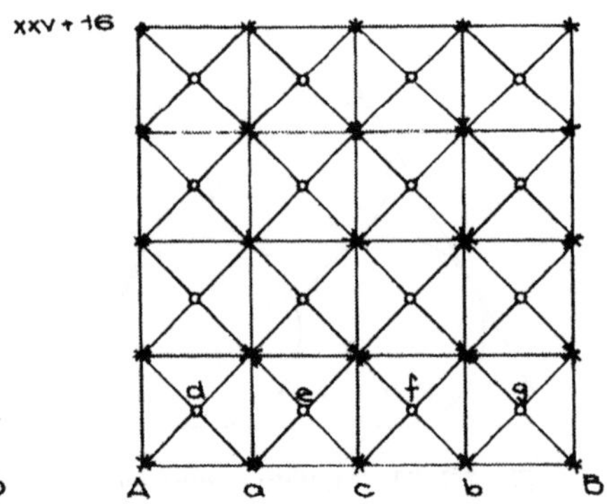

Fig. 1.4. Two examples by Farkas Bolyai for planting trees in a square-shaped region.

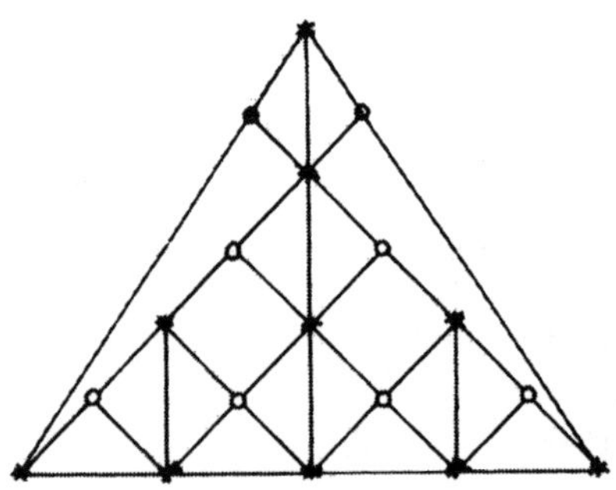

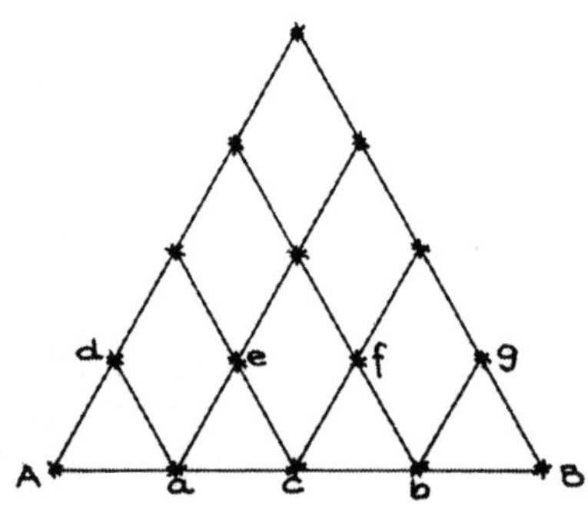

Fig. 1.5. Two examples by Farkas Bolyai for planting trees in a triangular region.

subject. Reference [113] gives more detailed information on this topic with further examples.

It should be mentioned here that the above packings of Farkas Bolyai are not necessarily optimal in terms of density. For instance, the packing of Figure 1.3 has been recently shown to be suboptimal: if the sides of the triangle are chosen to have unit length, the common radius of the circles in the packing of Bolyai is approximately 0.072169. However, in 1995, R. L. Graham and B. D. Lubachevsky discovered the packing configurations portrayed in Figure 1.6 with the approximate radii of 0.074360, 0.074326, and 0.074323, respectively [35]. (Recall that Bolyai did not intend to look for the densest packings, but actually investigated the convergence of the sequence of densities.)

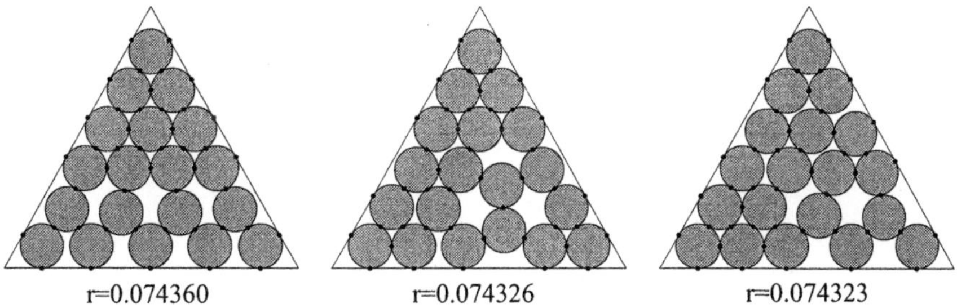

Fig. 1.6. The packings of 19 circles with densities larger than that of the example by Bolyai.

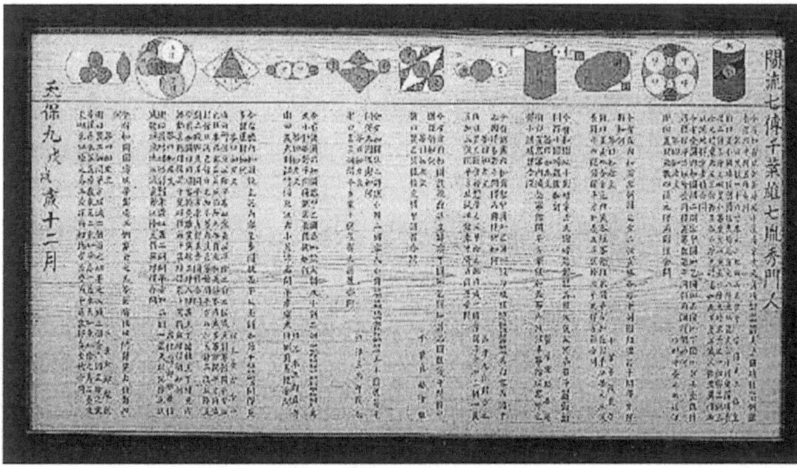

Fig. 1.7. A Japanese Sangaku.

1.2.3 Packing problems from ancient Japan

Historically, the work of Bolyai was not the very first on the packing of circles. There are other interesting early packings in fine arts, relics of religions, in nature [121], and also outside Europe. For instance, many of the Japanese 'Sangaku's of the Edo period (1603–1867) deal with various problems of locating circles in various context. The name Sangaku means mathematical tablet in English. These wooden tablets usually contain geometrical problems. They were displayed in temples and Shinto shrines, probably for meditation purposes (Figure 1.7).

Figure 1.8 shows an example where six equal circles have been packed in a rectangle, also from Japan. For more information on this interesting area of the history of mathematics, see [30], say.

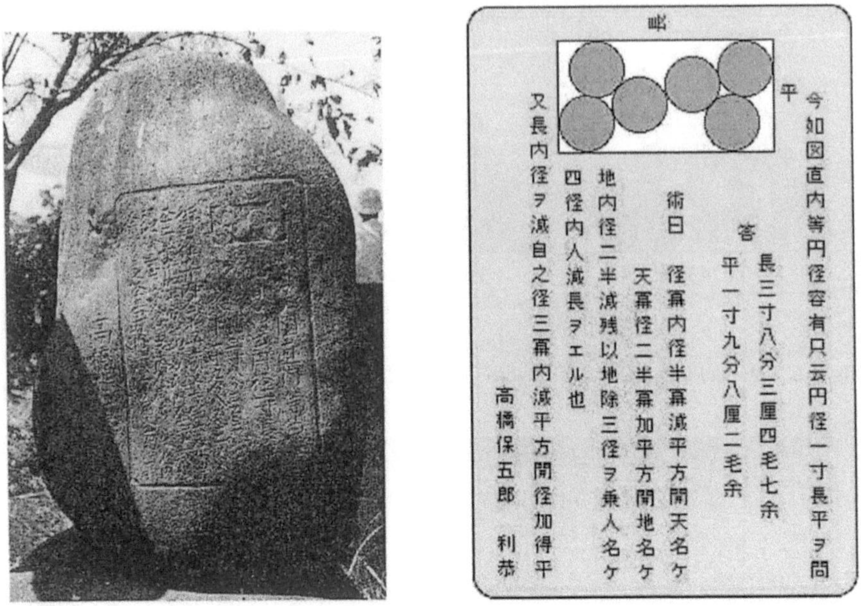

Fig. 1.8. The packing of six equal circles in a rectangle, carved on a rock in Japan.

1.2.4 A circle packing from Hieronymus Bosch

Perhaps the most extraordinary example depicting circle packings is the famous picture by the Dutch painter Hieronymus Bosch (c. 1450–1516): On the left wing of his triptych 'The Garden of Earthly Delights' (c. 1500), an arrangement of circles is displayed on the surface of a sphere. Interestingly, the packing is very hard to discover even in high-quality catalogues; the details become visible only after magnification (Figure 1.9).

1.2.5 The densest packing of circles in the plane

The first solution to the problem of determining the densest packing of equal circles in the plane was given by the Norwegian mathematician Axel Thue (1863–1922)—see [123, 124]. One can prove that the densest packing is the one coming from intuition: the hexagonal structure, where each circle is surrounded by six others (Figure 1.10). In 1773, J. L. Lagrange (1736–1813) proved that this structure results in the densest *lattice* packing on the plane [58]. The density of the hexagonal packing is $\pi/\sqrt{12}$, equal to the limit of the earlier mentioned packing sequence of Farkas Bolyai.

Note that the first optimality proof by Thue was not very convincing; it was completed by the Hungarian mathematician László Fejes Tóth (1915–2005) in 1940 [21].

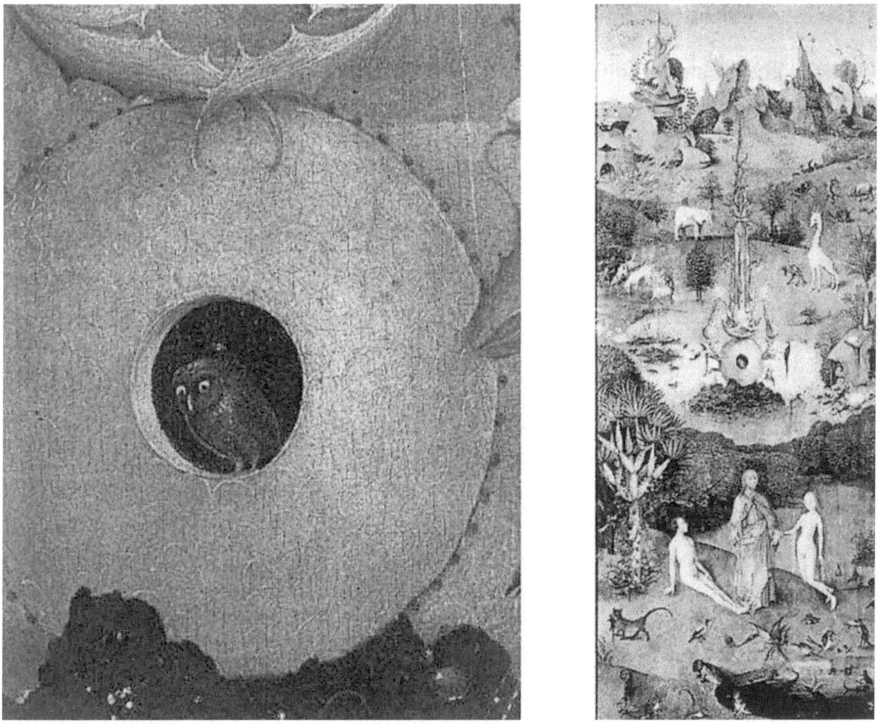

Fig. 1.9. Circles on a sphere in 'The Garden of Earthly Delights' by Hieronymus Bosch.

Fig. 1.10. The densest packing of identical circles in the plane.

1.2.6 Circle packings in bounded shapes

The literature of the studies related to the densest packing of identical circles in bounded regions is very rich. Besides the problem class of packing in a square, there are notable results for the packing of circles in a circle [26, 27, 28, 33, 37, 57, 76, 96, 101], in an equilateral triangle [35, 74, 77, 78, 91, 92], in an isosceles right triangle [45], and in a rectangle [103].

1.2.7 Generalizations and related problems

One can formulate the circle packing problems in such a way that the packing is carried out in some non-Euclidean geometry (e.g. in Bolyai-Lobachevsky geometry). However, in such cases even the proper way of defining the concept of density raises some difficulties [7]. Furthermore, the problem can be modified so that we can use a metric other than the usual Euclidean one [4, 22, 25]. Of course, one can consider packings in higher dimensions; see, for example the famous *Kepler Problem* [42], and the *Kissing Number* problem class [12].

1.2.8 Packing of equal circles in a unit square

The first appearance of this particular problem in the mathematical literature is the publication by Leo Moser [82] from 1960, in which he raised the following conjecture: *"eight points in or on a unit square determine a distance less than or equal to* $\frac{1}{2}\sec 15^o$*"*. (For historical completeness, however, note that from the memoirs of L. Fejes Tóth it turns out that he and his compatriot Dezső Lázár (1913– c. 1943) had already investigated the problem before 1940 [112, 117].)

The above conjecture of Moser was proved by J. Schaer and A. Meir [105] in 1965. In the latter publication the authors mentioned that for $n = 6$ there already exists a proof by R. L. Graham, and they were also aware of a proof for $n = 7$. In the end, neither of these claimed proofs was published. (The optimal arrangements are obvious up to $n = 5$.) Schaer sent his early proof for $n = 7$ to P. G. Szabó (one of the authors of this volume), adding a comment that he thought it was an 'ugly proof', so he had decided not to publish it. To the best of our knowledge, no optimality proof for the case $n = 7$ has yet been published (with purely mathematical tools, i.e. without using computers). The $n = 6$ case was later solved by B. L. Schwartz [108] in 1970 and by H. Melissen [75] in 1994, independently of each other.

In 1965, Schaer solved the $n = 9$ case as well [103]. In 1970, M. Goldberg reviewed the previous results [32] and gave conjecturally optimal packings up to $n = 27$ and for some further number of circles. Between 1970 and 1990, at least ten papers reported solutions for $n = 10$ [32, 38, 39, 40, 79, 80, 95, 104, 107, 126]. Table 1.1 gives an overview of the improvements of the best-known packings until 1990, when C. de Groot, R. Peikert, and D. Würtz found the densest packing and proved its optimality with a computer-assisted method [39]. In the table, r_{10} denotes the common radius of the circles (given to six decimal places) packed in the unit square.

A conventional proof for the optimality for ten circles still does not exist, although there are many promising results in this direction as well – see, for example, the work of M. Hujter [49].

Based on their computer technique, in 1992 R. Peikert, D. Würtz, M. Monagan, and C. de Groot published the densest circle packings up to $n = 20$ [94].

In 1983, G. Wengerodt reported an optimality proof for $n = 16$ [127], using only conventional mathematical tools. In 1987, he extended his results for the optimality of $n = 14$ [128], $n = 25$ [129], and (together with K. Kirchner) for $n = 36$ [53]. However, recently there have been some doubts raised concerning the correctness of

Table 1.1. Results for the packing of 10 circles.

Year	Author(s)	r_{10}	Reference
1970	M. Goldberg	0.147060	[32]
1971	J. Schaer	0.147770	[104]
1979	K. Schlüter	0.148204	[107]
1987	R. Milano	0.147920	[79]
1989	G. Valette	0.148180	[126]
1990	B. Grünbaum	0.148197	[40]
1990	M. Grannell	0.148204	[38]
1990	M. Mollard and C. Payan	0.148204	[80]
1990	C. de Groot, R. Peikert, and D. Würtz	0.148204	[39]

the proofs for $n = 25, 36$; for details, see the notes of R. Blind in the Mathematical Reviews (MR1453444).

In 1995, C. D. Maranas, C. A. Floudas, and P. M. Pardalos reported packings for up to 30 circles [66] (without proving their optimality), using the nonlinear programming solver MINOS. Later it turned out that the published result for $n = 21$ was incorrect: namely, the reported function value was better than that of the proven optimum obtained a couple of years later. This also indicates that special care must be taken (that is, numerically reliable calculations must be used) when computer techniques are applied to find approximate or optimal packings.

In 1996, T. Tarnai and Zs. Gáspár [122] improved the packing configuration by M. Goldberg [32] for $n = 19$ (independently of the results of Peikert et al.) and reported upper bounds for the packing densities.

In 1997, K. J. Nurmela and P. R. J. Östergård published approximately optimal packings up to $n = 50$ [86]. In their work, some new classes of packing patterns were introduced. These classes were extended by R. L. Graham and B. D. Lubachevsky [36] with the aid of new results obtained by a method called *billiard simulation*.

In 1999, K. J. Nurmela and P. R. J. Östergård gave computer-assisted optimality proofs for the cases $n \leq 27$ [87]. Together with R. aus dem Spring, they published new theoretical lower bounds for the optimum values, and determined a conjecturally optimal packing sequence [89].

In 2000, D. W. Boll, J. Donovan, R. L. Graham, and B. D. Lubachevsky improved the best-known packings for $n = 32, 37, 48, 50$ [5]. In the same year, P. Ament and G. Blind claimed an improved packing for $n = 34$ (although at that time a better solution was already known for that case) [3].

In 2002, M. Locatelli and U. Raber developed a deterministic computer method, which enabled them to improve the existing best packings for $n = 32$ and 37, and to prove the approximate optimality of the best existing packings for $n = 10 - 35, 38$ and 39 [61].

In 2005, B. Addis, M. Locatelli, and F. Schoen further improved the previously known best packings for $n = 53, 59, 66, 68, 73, 77, 85, 86$ [1].

Overall, then, the currently available results for the problem class are the following:

Table 1.2. The authors of the known optimal packings.

Year	Authors	Results for n
1965	J. Schaer and A. Meir [103, 105]	$8, 9$
1970	B. L. Schwartz [108]	6
1983	G. Wengerodt [127]	16
1987	G. Wengerodt [128]	14
1992	R. Peikert et al. [94]	$10 - 20$
1999	K. J. Nurmela and P. R. J. Östergård [87]	$7, 21 - 27$
2004	M. Cs. Markót [68]	28
2005	M. Cs. Markót and T. Csendes [70]	$29 - 30$

– *Proven (optimal) packings* are known up to $n = 30$. The cases $n = 2, \ldots, 6, 8, 9$, 14 and 16 are solved by conventional mathematical tools, while the others are results of computer-assisted methods. Table 1.2 summarizes the known optimal packings and the corresponding publications.

– *Approximate packings* (i.e. packings determined by various methods without proof of optimality) are reported in the literature for up to 200 circles. Table 1.3 contains the most important improvements of the last decade. These numerical results have been obtained via several different strategies; for instance, using billiard simulation [36, 62, 65], the minimization of an energy function [86], standard BFGS quasi-Newton algorithm [94], a nonlinear programming solver (MINOS 5.3) [66], simulated annealing and threshold accepting techniques [11], and the Cabri-Géomètre software [80]. Although it is not known whether these solutions are optimal, in some cases the numerical results can help us find better solutions when they are used as lower bounds of the optimal (maximal) solutions. For instance, several strategies described in the packing literature, like that of R. Peikert et al. in [94], are based on the knowledge of a good lower bound of the optimum value.

– Finally, numerous other *related results* (concerning e.g. patterns, bounds, and various properties of the optimal solutions) have been published as well [36, 61, 89, 114, 118].

Some other details on the history of the problem of packing equal circles in a square (PECS) can be found in [8, 9, 66, 94, 118, 120, 130].

Table 1.3. The authors of approximate packings between 1995 and 2006.

Year	Authors	Results for n
1995	C. D. Maranas et al. [66]	up to 30
1996	R. L. Graham and B. D. Lubachevsky [36]	up to 61
1997	K. J. Nurmela and P. R. J. Östergård [86]	up to 50
2000	D. W. Boll et al. [5]	$32, 37, 48, 50$
2001	L. G. Casado et al. [11]	up to 100
2002	M. Locatelli and U. Raber [61]	up to 39
2005	B. Addis et al. [1]	50 – 100
2006	P. G. Szabó and E. Specht [120]	up to 200

2

Problem Definitions and Formulations

In this chapter we will specify the densest packing of equal circles in a square problem, and discuss some equivalent problem settings. Since, besides the geometric investigations, we also consider the problem from a global optimization point of view, some possible mathematical programming models will be included here.

2.1 Geometrical models

Informally speaking, the packing circles in a square and its related problems can be described in the following ways:

Problem 2.1. Place $n \geq 2$ equal and non-overlapping circles in a square, such that the common radius of the circles is maximal.

Problem 2.2. Place $n \geq 2$ points in a square, such that the minimum of the pairwise distances is maximal.

Problem 2.3. Place $n \geq 2$ equal and non-overlapping circles with the common radius in the smallest possible square.

Problem 2.4. Place $n \geq 2$ points with pairwise distances of at least a given positive value in the smallest possible square.

Of course, in order to investigate these problems and their relations in detail, we need their formal definitions and a consistent system of notation.

Formal description of Problem 2.1:

Definition 2.5. $P(r_n, S) \in P_{r_n}$ *is a* circle packing *with the common radius* r_n *in the square* $[0, S]^2$, *where* $P_{r_n} = \{((x_1, y_1), \ldots, (x_n, y_n)) \in [0, S]^{2n} \mid (x_i - x_j)^2 + (y_i - y_j)^2 \geq 4r_n^2; x_i, y_i \in [r_n, S - r_n]\ (1 \leq i < j \leq n)\}$. $P(r_n, S) \in P_{\overline{r}_n}$ *is an* optimal circle packing, *if* $\overline{r}_n = \max\limits_{P_{r_n} \neq \emptyset} r_n$.

Problem $\mathcal{P}_1^n$. Determine the optimal circle packings for a given $n \geq 2$ integer.

Formal description of Problem 2.2:

Definition 2.6. $A(m_n, \Sigma) \in A_{m_n}$ *is a* point arrangement *with the minimal pairwise distance* m_n *in the square* $[0, \Sigma]^2$, *where* $A_{m_n} = \{((x_1, y_1), \ldots, (x_n, y_n)) \in [0, \Sigma]^{2n} \mid (x_i - x_j)^2 + (y_i - y_j)^2 \geq m_n^2; (1 \leq i < j \leq n)\}$. $A(m_n, \Sigma) \in A_{\overline{m}_n}$ *is an* optimal point arrangement , *if* $\overline{m}_n = \max\limits_{A_{m_n} \neq \emptyset} m_n$.

Problem $\mathcal{P}_2^n$. Determine the optimal point arrangements for a given $n \geq 2$ integer.

Formal description of Problem 2.3:

Definition 2.7. $P'(R, s_n) \in P'_{s_n}$ *is an* associated circle packing *with the common radius* R *in the square* $[0, s_n]^2$, *where* $P'_{s_n} = \{((x_1, y_1), \ldots, (x_n, y_n)) \in [0, s_n]^{2n} \mid (x_i - x_j)^2 + (y_i - y_j)^2 \geq 4R^2; x_i, y_i \in [R, s_n - R]\ (1 \leq i < j \leq n)\}$. $P'(R, s_n) \in P'_{\overline{s}_n}$ *is an* optimal associated circle packing , *if* $\overline{s}_n = \min\limits_{P'_{s_n} \neq \emptyset} s_n$.

Problem $\mathcal{P}_3^n$. Determine the optimal associated circle packings for a given $n \geq 2$ integer.

Formal description of Problem 2.4:

Definition 2.8. $A'(M, \sigma_n) \in A'_{\sigma_n}$ *is an* associated point arrangement *with the minimal pairwise distance* M *in the square* $[0, \sigma_n]^2$, *where* $A'_{\sigma_n} = \{((x_1, y_1), \ldots, (x_n, y_n)) \in [0, \sigma_n]^2 \mid (x_i - x_j)^2 + (y_i - y_j)^2 \geq M^2\ (1 \leq i < j \leq n)\}$. $A'(M, \sigma_n) \in A'_{\overline{\sigma}_n}$ *is an* optimal associated point arrangement, *if* $\overline{\sigma}_n = \min\limits_{A'_{\sigma_n} \neq \emptyset} \sigma_n$.

Problem $\mathcal{P}_4^n$. Determine the optimal associated point arrangements for a given $n \geq 2$ integer.

Theorem 2.9. *The Problems $\mathcal{P}_1^n$, $\mathcal{P}_2^n$, $\mathcal{P}_3^n$, and $\mathcal{P}_4^n$ are equivalent in the sense that, if someone is able to solve one of the problem types for a fixed $n \geq 2$ integer, then this solution yields the solutions of all the other problem types. That is, for each n the optimal solutions of the particular problems can be derived from each other.*

The theorem will be proved through four lemmas, each of them giving the equivalence of two different problems. First, we prove that the circle centres of a $P(r_n, S)$ optimal circle packing result in an $A(m_n, \Sigma)$ optimal point arrangement. Then we show that the circles drawn around the points of an $A(m_n, \Sigma)$ optimal point arrangement with a proper radius result in a $P'(R, s_n)$ optimal associated circle packing. In the next step, we prove that circle centres of a $P'(R, s_n)$ optimal associated circle packing result in an $A'(M, \sigma_n)$ optimal associated point arrangement. Then we show that the circles drawn around the points of an $A'(M, \sigma_n)$ optimal associated point arrangement with a proper radius result in a $P(r_n, S)$ optimal circle packing.

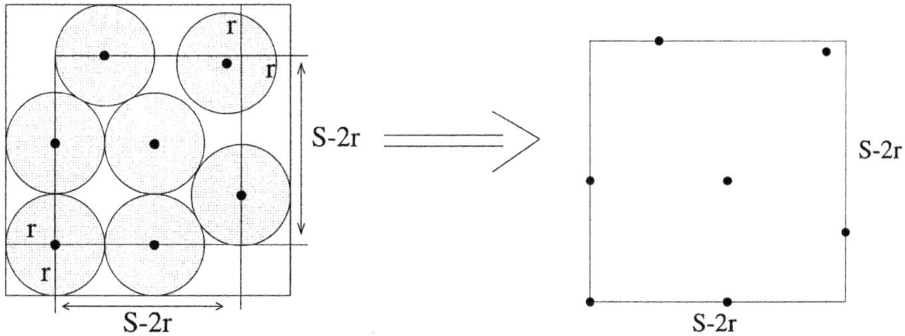

Fig. 2.1. Each circle packing corresponds to a point arrangement.

Lemma 2.10. *Let $P(r_n, S)$ be an optimal circle packing. Then the centres of the circles correspond to an $A(m_n, \Sigma)$ optimal point arrangement with $\Sigma = S - 2r_n$, $m_n = 2r_n$.*

PROOF. Assume the opposite of the statement of the lemma, i.e. that $A(2r_n, S - 2r_n)$ is not optimal. Then there exists an $A(m'_n, S - 2r_n)$ point arrangement with $m'_n > 2r_n$. Create a circle packing from $A(m'_n, S - 2r_n)$ by drawing a circle of radius $m'_n/2$ from each point of the packing (see Figure 2.1). Clearly, each such circle is located in a square of side $S - 2r_n + m'_n$. Reduce this square to a square of side S, using the midpoint of the square as the centre of the transformation. The transformation changes the common radius of the circles to

$$\frac{m'_n S}{2(S - 2r_n + m'_n)}.$$

$P(r_n, S)$ is an optimal circle packing, thus

$$r_n \geq \frac{m'_n S}{2(S - 2r_n + m'_n)}$$

holds, from which we get $(2r_n - m')(S - 2r_n) \geq 0$. Since it is easy to see that $S - 2r_n > 0$ holds for all $n \geq 2$, we obtain $2r_n - m' \geq 0$, but this contradicts the original assumption that $m' > 2r_n$. This contradiction completes the proof. □

Lemma 2.11. *Let $A(m_n, \Sigma)$ be an optimal point arrangement. Then the circles drawn around the points with a common radius of $R = m_n/2$ form a $P'(R, s_n)$ optimal associated circle packing with $s_n = \Sigma + m_n$.*

PROOF. Assume that the $P'(m_n/2, \Sigma + m_n)$ associated circle packing is not optimal. Then there exists a $P'(m_n/2, s'_n)$ associated circle packing with $s'_n < \Sigma + m_n$. The centres of this latter packing are located in a square of side $s'_n - m_n$. Enlarge this square to a square of side Σ, using the midpoint of the square as the centre of the transformation. The enlargement changes the minimal pairwise distance between the circle centres to

$$\frac{m_n \Sigma}{s'_n - m_n}.$$

Since $A(m_n, \Sigma)$ is an optimal point arrangement,

$$m_n \geq \frac{m_n \Sigma}{s'_n - m_n}$$

holds. Using the obvious $m_n > 0$ condition, we can divide both sides by m_n, and after rearrangement we obtain $s'_n - m_n \geq \Sigma$. But this contradicts the assumption $s'_n < \Sigma + m_n$. □

Lemma 2.12. *Let $P'(R, s_n)$ be an optimal associated circle packing. Then the centres of the circles correspond to an $A'(M, \sigma_n)$ optimal associated point arrangement with $M = 2R, \sigma_n = s_n - 2R$.*

PROOF. Assume that $A'(2R, s_n - 2R)$ is not optimal. Then there exists an $A'(2R, \sigma'_n)$ associated point arrangement for which $\sigma'_n < s_n - 2R$. Create a circle packing from this latter arrangement by drawing a circle of radius R around each point. These circles are located in a square of side $\sigma'_n + 2R$. Since $P'(R, s_n)$ is an optimal associated circle packing, $s_n \leq \sigma'_n + 2R$, which contradicts our original assumption $\sigma'_n < s_n - 2R$. □

Lemma 2.13. *Let $A'(M, \sigma_n)$ be an optimal associated point arrangement. Then the circles drawn around the points with a common radius of $r_n = M/2$ form a $P(r_n, S)$ optimal circle packing with $S = \sigma_n + M$.*

PROOF. Assume that the $P(M/2, \sigma_n + M)$ circle packing is not optimal, that is, there exists a $P(r'_n, \sigma_n + M)$ circle packing with $r'_n > M/2$. The centres of this latter packing are located in a square of side $\sigma_n + M - 2r'_n$. Reduce this square (using the midpoint of the square as the centre of the transformation) in such a way that the minimal pairwise distance between the centres becomes M. Then the side of the reduced square is

$$\frac{M(\sigma_n + M - 2r'_n)}{2r'_n}.$$

Since $A'(M, \sigma_n)$ is an optimal associated point arrangement, we obtain

$$\sigma_n \leq \frac{M(\sigma_n + M - 2r'_n)}{2r'_n},$$

from which, after rearrangement, we get $0 \leq (\sigma_n + M)(M - 2r'_n)$. The first term on the right-hand side is positive, which yields $M - 2r'_n \geq 0$, but this contradicts the assumption $r'_n > M/2$. □

Table 2.1. Relations between the parameters of the problems.

	$P(r_n, S)$	$A(m_n, \Sigma)$
$P(r_n, S)$	1	$r_n = \frac{Sm_n}{2(m_n+\Sigma)}$
$A(m_n, \Sigma)$	$m_n = \frac{2\Sigma r_n}{S-2r_n}$	1
$P'(R, s_n)$	$s_n = \frac{RS}{r_n}$	$s_n = \frac{2R(m_n+\Sigma)}{m_n}$
$A'(M, \sigma_n)$	$\sigma_n = \frac{M(S-2r_n)}{2r_n}$	$\sigma_n = \frac{M\Sigma}{m_n}$

Table 2.2. Relations between the parameters of the problems.

	$P'(R, s_n)$	$A'(M, \sigma_n)$
$P(r_n, S)$	$r_n = \frac{RS}{s_n}$	$r_n = \frac{MS}{2(M+\sigma_n)}$
$A(m_n, \Sigma)$	$m_n = \frac{2R\Sigma}{s_n-2R}$	$m_n = \frac{M\Sigma}{\sigma_n}$
$P'(R, s_n)$	1	$s_n = \frac{2R(M+\sigma_n)}{M}$
$A'(M, \sigma_n)$	$\sigma_n = \frac{M(s_n-2R)}{2R}$	1

Moreover, of course, the transformed circle packings and point arrangements used in the previous proofs are in shifted squares, where the left lower corner of a square is in the (0,0) point.

Definition 2.14. *The density of a circle packing* $P(r_n, S)$ *is given by the formula*

$$d_n(r_n, S) = \frac{nr_n^2\pi}{S^2}.$$

Since the density of a circle packing is a quadratic function of the radius r_n, the following problem formulation is equivalent (in the sense of Theorem 2.9) to the Problems $\mathcal{P}_i^n$, $1 \le i \le 4$:

Problem $\mathcal{P}_5^n$. Determine the *densest* $P(r_n, S)$ circle packings for a given $n \ge 2$ integer.

Corollary 2.15. *The relations between the parameters of the circle packings, point arrangements, associated circle packings, and associated point arrangements* derived from each other *are the ones listed in Tables 2.1 and 2.2.*

PROOF. Each entry of Tables 2.1 and 2.2 is a straightforward consequence of the transformations used in the proofs of Lemmas 2.10 to 2.13. □

In the sequel, the Problems $\mathcal{P}_1^n$, $\mathcal{P}_2^n$, and $\mathcal{P}_5^n$ will be investigated *in the unit square.* (Obviously, one can fix the square this way without the loss of generality, since the different domain squares can be transformed to each other without affecting the structures of the packings.)

In the following, we consider two packings to be *identical*, if they can be transformed to each other by applying symmetry transformations or index permutations. Note that this consideration is obvious from a *geometrical point of view*. However, as we shall see later, one of the main difficulties the numerical methods have to cope with is finding one configuration and avoiding the other identical configurations.

Definition 2.16. *Let us suppose that there is a given solution of Problem (2.1) (or the equivalent ones (2.2)–(2.4)). We say that* a circle is free (or a rattler) *if its centre can be moved towards a positive distance point without causing the others overlap.*

Definition 2.17. *We say, that*

- *a circle is* fixed *if it isn't a free circle,*
- *a packing is* rigid *if all of its circles are fixed.*

We should point out here that when a packing contains one or more free circles, then the solution is obviously not unique. Moreover, the possible locations of the centre of any free circles form a non-empty interior and connected set. In the present volume the number of contacts will be denoted by c_n, and the number of free circles by f_n. In each figure a contact will be represented by a short line section and free circles will be indicated by dark shading.

2.2 Mathematical programming models

As we have already mentioned, the circle packing problem was originally a geometrical problem; on the other hand, it can also be viewed as a continuous global optimization problem.

In this book, we will usually refer to the following *bound-constrained, max-min optimization model* of the point arrangement problem in the unit square (that is, Problem 2.2 or Problem $\mathcal{P}_2^n$):

$$\begin{aligned} &\max_{x_i,y_i} \quad \min_{1\le i<j\le n} \sqrt{(x_i-x_j)^2+(y_i-y_j)^2} \\ &\text{subject to} \quad 0\le x_i,y_i\le 1 \qquad (1\le i\le n), \end{aligned} \tag{2.1}$$

where x_i, y_i are the coordinates of the i-th point. The goals are to find the global optimum of the problem (the maximum of the minimal pairwise distance of the points), *and also*, to find the global optimizer(s), that is, the respective locations of the points.

Besides (2.1), the mathematical programming models of Problem $\mathcal{P}_2^n$ can be represented in various different ways, for instance:

a) *as a continuous, nonlinear, inequality-constrained global optimization problem* [66]:

$$\max_{x_i, y_i} \ t,$$

subject to

$$\sqrt{(x_i - x_j)^2 - (y_i - y_j)^2} \geq t \qquad (1 \leq i < j \leq n),$$

$$0 \leq x_i, y_i \leq 1 \qquad (1 \leq i \leq n).$$

b) *as a DC programming problem* [48]:

A DC (difference of convex functions) programming problem is a mathematical programming problem, where the objective function is given by the difference between two convex functions. In [48], the *minimal pairwise squared distance* is treated as the objective function, and a DC decomposition of this function is given. That is, the point arrangement problem is formulated as

$$\max_{\substack{z_j \in [0,1]^2, \\ 1 \leq j \leq 2n}} \ \min_{1 \leq i < k \leq n} \left\{ (z_i - z_k)^2 + (z_{n+i} - z_{n+k})^2 \right\},$$

with

$$z = (x_1, \ldots, x_n, y_1, \ldots, y_n),$$

which is obviously equivalent to the original Problem $\mathcal{P}_2^n$. Using the additional notation

$$J = \{1, \ldots, 2n\},$$
$$J_{ik} = \{i, k, n+i, n+k\},$$

the objective function can be written as

$$\min_{1 \leq i < k \leq n} \left\{ (z_i - z_k)^2 + (z_{n+i} - z_{n+k})^2 \right\}$$

$$= \min_{1 \leq i < k \leq n} \left\{ (z_i - z_k)^2 + (z_{n+i} - z_{n+k})^2 - 2 \sum_{j=1}^{2n} z_j^2 + 2 \sum_{j=1}^{2n} z_j^2 \right\}$$

$$= 2 \sum_{j=1}^{2n} z_j^2 + \min_{1 \leq i < k \leq n} \left\{ -2 \sum_{j \in J \setminus J_{ik}} z_j^2 - (z_i + z_k)^2 - (z_{n+i} + z_{n+k})^2 \right\}$$

$$= 2 \sum_{j=1}^{2n} z_j^2 - \max_{1 \leq i < k \leq n} \left\{ 2 \sum_{j \in J \setminus J_{ik}} z_j^2 + (z_i + z_k)^2 + (z_{n+i} + z_{n+k})^2 \right\}.$$

Thus, the objective function can be specified as the difference of two *convex* functions $g : \mathbb{R}^{2n} \to \mathbb{R}^+$ and $h : \mathbb{R}^{2n} \to \mathbb{R}^+$:

$$g(z) = 2\sum_{j=1}^{2n} z_j^2,$$

$$h(z) = \max\left\{\left(2\sum_{j\in J\setminus J_{ik}} z_j^2 + (z_i + z_k)^2 + (z_{n+i} + z_{n+k})^2\right) : 1 \le i < k \le n\right\}.$$

c) *as an all-quadratic optimization problem (or QCQP – Quadratically Constrained Quadratic Problem)*:

The general form of an all-quadratic optimization problem [97] is

$$\min \ \left[x^T Q^0 x + (d^0)^T x\right],$$

subject to

$$x^T Q^l x + (d^l)^T x + c^l \le 0 \quad l = 1, \ldots, p$$

$$x \in P,$$

where Q^l $(l = 0, \ldots, p)$ are real $(n+1) \times (n+1)$-dimensional matrices, d^l $(l = 0, \ldots, p)$ are real $(n+1)$-dimensional vectors, c^l $(l = 1, \ldots, p)$ are real numbers, p is the number of constraints, and P is a polyhedron.

Problem $\mathcal{P}_2^n$ can be written as a special case of the all-quadratic optimization problem with a linear objective function in the following way:

$$Q^0 = \mathbf{0}, \quad x^T = (x_0, x_1, \ldots, x_{2n}), \quad (d^0)^T = (-1, 0, \ldots, 0),$$

$$(d^l)^T = \underline{0}, \ c^l = 0, \ p = \frac{n(n-1)}{2}, \quad P = [0, \sqrt{2}] \times [0, 1]^{2n},$$

$$[Q^l]_{ij} = Q_{ij}^{l'l''} = \begin{cases} -1, & \text{if } i = j = \begin{cases} 2l', \\ 2l'', \\ 2l'+1, \\ 2l''+1, \end{cases} \\ 1, & \text{if } i = j = 1, \\ & \quad i = 2l''+1 \ \text{ and } j = 2l'+1, \\ & \quad i = 2l'' \qquad \text{and } j = 2l', \\ & \quad i = 2l'+1 \ \text{ and } j = 2l''+1, \\ & \quad i = 2l' \qquad \text{and } j = 2l'', \\ 0, & \text{otherwise}, \end{cases}$$

$$1 \le i, j \le 2n+1,$$

$$1 \le l' < l'' \le n.$$

In this model, x_0 is the minimal distance between the points. The coordinates of the i-th point $(1 \le i \le n)$ are (x_{2i-1}, x_{2i}).

The above models may be of interest for mathematical programming solvers as very hard optimization problems. To demonstrate the expected difficulty of the problem, we might mention the example of the sophisticated, interval arithmetic-based global optimization solver GlobSol [51], which was unable to return sufficient results even for the case of packing five circles/points with reasonable parameter settings [125]. However, as the previous and current numerical studies show, approaches that use not only optimization models, but also the geometrical aspects of the problem are often more effective (cf. the methods of Chapters 7 and 8). Hence in the next chapter some of these useful geometrical characteristics will be investigated in detail.

3

Bounds for the Optimum Values

In this chapter, we give lower and upper bounds for the maximal radii $\overline{r}_n$ of the circle packing problems (Problem 2.1 or $\mathcal{P}_1^n$), and for the maximal distances $\overline{m}_n$ of the point arrangement problems (Problem 2.2 or $\mathcal{P}_1^n$) in the unit square. The bounds obtained are valid in general, i.e. they are applicable for arbitrary $n \geq 2$ values. The error term of the asymptotic relation has also been approximated.

In Section 3.1, a lower bound for the maximal distance $\overline{m}_n$ will be given based on a regular arrangement. In Section 3.2, upper bounds for the maximal radii $\overline{r}_n$ will be determined. The latter bounds are based on computation of the minimum 'gaps' occurring between the circles and between the circles and sides of the square.

3.1 Lower bounds for the maximal distance $\overline{m}_n$

The construction of the lower bounds for the optimum of (2.1) is based on the simple fact, that the minimal distance value for an *arbitrary feasible* solution can be used as a lower bound of the maximal distances $\overline{m}_n$. Notice that for studies on the lower bounds it is natural to deal with the point arrangement problem since feasible point arrangements can easily be generated. The only constraint is that the points must be located in the unit square (in contrast to the circle packing problem, where it is much harder to find feasible solutions because of the non-overlapping restriction).

The following theorem provides a lower bound for $\overline{m}_n$ with arbitrary n values:

Theorem 3.1. *For each $n \geq 2$,*

$$\sqrt{\frac{2}{\sqrt{3}n}} < \overline{m}_n. \tag{3.1}$$

PROOF. First we will show that it is enough to prove that it is possible to locate at least $\frac{2}{\sqrt{3}}\sigma^2$ points in a square of side σ. Then the statement (3.1) will be valid.

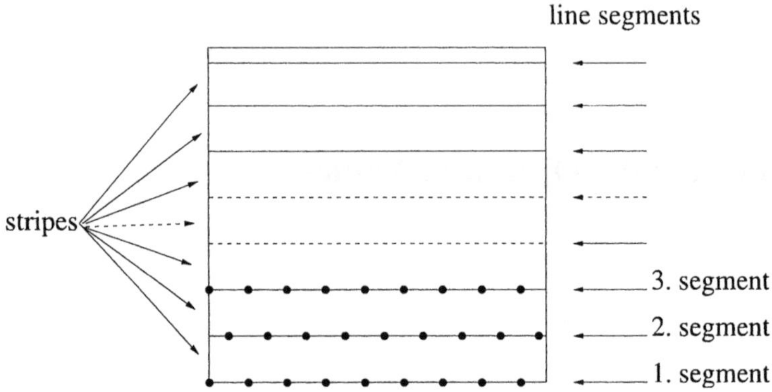

Fig. 3.1. Subdividing the square into strips.

Indeed, from Theorem 2.9 and from Corollary 2.15, $A'(1,\sigma)$ corresponds to an $A(m,1)$ point arrangement with $\sigma m = 1$. Thus,

$$\frac{2}{\sqrt{3}m^2} = \frac{2}{\sqrt{3}\frac{1}{\sigma^2}} = \frac{2}{\sqrt{3}}\sigma^2 < n.$$

But this implies (3.1), since

$$\frac{2}{\sqrt{3}n} < m^2 \leq \overline{m}_n^2.$$

We prove the existence of the required associated point arrangement in a constructive way: First, subdivide the square into equal 'stripes' of width $\frac{\sqrt{3}}{2}$, starting from one of the sides of the square of side σ, as is shown in Figure 3.1. (Of course the last stripe may have smaller width.) These equal stripes are determined by $\left\lfloor \frac{2\sigma}{\sqrt{3}} \right\rfloor + 1$ parallel line segments in the square. Place $\lfloor \sigma \rfloor + 1$ equidistant points on the first line segment in such a way that the first point goes to the corner of the square, and the distance between the neighboring points is 1. On the second line segment, place $\lfloor \sigma - \frac{1}{2} \rfloor + 1$ equidistant points (again, with the distances of 1 to the neighboring points) in such a way that the first one is at a distance of 1/2 from the end point of the line segment.

For the further line segments, alternate the above two ways of placing points by locating $\lfloor \sigma \rfloor + 1$ points on each line segment with an odd index, and locating $\lfloor \sigma - \frac{1}{2} \rfloor + 1$ points on each line segment with an even index. In this way the distance between each pair of points will be at least 1. Thus, the total number of points placed in the square is

$$\frac{\left\lfloor \frac{2\sigma}{\sqrt{3}} \right\rfloor + 1}{2}\left(\lfloor \sigma \rfloor + \left\lfloor \sigma - \frac{1}{2} \right\rfloor + 2\right), \quad \text{if } \left\lfloor \frac{2\sigma}{\sqrt{3}} \right\rfloor + 1 \text{ is even,}$$

and

$$\frac{\left\lfloor \frac{2\sigma}{\sqrt{3}} \right\rfloor}{2}\left(\left\lfloor \sigma - \frac{1}{2} \right\rfloor + \lfloor \sigma \rfloor + 2\right) + \lfloor \sigma \rfloor + 1, \quad \text{if } \left\lfloor \frac{2\sigma}{\sqrt{3}} \right\rfloor + 1 \text{ is odd.}$$

In both cases, it is easy to see that we have placed at least $\frac{2}{\sqrt{3}}\sigma^2$ points in total, as because of $\lfloor \sigma - \frac{1}{2} \rfloor + \lfloor \sigma \rfloor + 2 > 2\sigma$, $\left\lfloor \frac{2\sigma}{\sqrt{3}} \right\rfloor > \frac{2}{\sqrt{3}}\sigma - 1$, and $\lfloor \sigma \rfloor > \sigma - 1$, we get

$$\frac{\left\lfloor \frac{2\sigma}{\sqrt{3}} \right\rfloor + 1}{2}\left(\lfloor \sigma \rfloor + \left\lfloor \sigma - \frac{1}{2} \right\rfloor + 2\right) > \frac{\frac{2\sigma}{\sqrt{3}} - 1 + 1}{2} 2\sigma = \frac{2}{\sqrt{3}}\sigma^2$$

in the even case, and

$$\frac{\left\lfloor \frac{2\sigma}{\sqrt{3}} \right\rfloor}{2}\left(\left\lfloor \sigma - \frac{1}{2} \right\rfloor + \lfloor \sigma \rfloor + 2\right) + \lfloor \sigma \rfloor + 1 > \frac{\frac{2\sigma}{\sqrt{3}} - 1}{2} 2\sigma + \sigma - 1 + 1 = \frac{2}{\sqrt{3}}\sigma^2$$

in the odd case, respectively. □

Theorem 3.1 supplies lower bounds of general validity; that is, the inequality (3.1) is applicable for each $n \geq 2$ integer. For individual n values, computer methods, say, can be used to deliver good lower bounds for the optimum values—provided that either a geometric construction with exact values is derived from the numerical solutions, or the mathematical correctness of the numerical bounds is *validated* (verified) by some guaranteed reliability numerical method. A stochastic algorithm for finding approximate numerical solutions of the point arrangement problems is introduced in Chapter 4.

Besides this, it is known that for several values of n, the optimal solutions are regular arrangements. Keeping this observation in mind, one can speculate that other types of regular arrangements may also provide good lower bounds for the optimum values of Problem 2.2 (see also as (2.1)). Indeed, a detailed investigation of these special patterns resulted in further improvements on the theoretical lower bounds for several n values. These results will be discussed in Chapter 10.

3.2 Upper bounds for the optimal radius $\overline{r}_n$

As one might expect, it is somewhat harder to determine non-trivial upper bounds for $\overline{m}_n$ and $\overline{r}_n$ than those for lower bounds. The main difficulty for this is that lower bounds can be obtained in a constructive way by trying out proper approximate (feasible) solutions, but for the upper bounds some knowledge about the properties of the optimal solutions is required beforehand.

As we already mentioned in Subsection 1.2.5, in [21] it was proved that the densest packing of circles in the plane is a regular (hexagonal) arrangement, and the density of this packing is $\pi/\sqrt{12} \approx 0.9068996821$. Moreover, it can be proved that this value is an upper bound for the density of any circle packings in the unit square (see Chapter 11). However, when Problem 2.2 itself is investigated, this general upper bound can be substantially improved:

Theorem 3.2. *For each $n \geq 2$ integer, an upper bound of the $\overline{r}_n$ optimal radius is*

$$\overline{r}_n \leq \min\left(U_1(n), U_2(n)\right),$$

with

$$U_1(n) := \frac{1}{\sqrt{2\sqrt{3}n + (4\lfloor\sqrt{n}\rfloor - 2)(2-\sqrt{3})}},$$

and

$$U_2(n) := \frac{1+\sqrt{1+\frac{2}{\sqrt{3}}(n-1)}}{2n+2\sqrt{1+\frac{2}{\sqrt{3}}(n-1)}},$$

where $\lfloor\cdot\rfloor$ denotes the greatest integer not greater than the argument.

PROOF. The proof is in two parts. First we show that $U_1(n)$ is the upper bound of $\overline{r}_n$, then we do likewise with $U_2(n)$.

- $U_1(n)$: Suppose that for a fixed $n \geq 2$ an optimal circle packing in the unit square $[0,1]^2$ exists, and that the common radius of the circles is $\overline{r}_n$. Partition the square into *Dirichlet cells* (*Voronoi polygons*) according to the centres of the circles: the interior of each cell consists of exactly those points of the square which are closer to a circle centre than to any other. Then, each cell belonging to a circle centre will contain the respective circle itself as well. Obviously the total area of the n cells will be equal to 1.

  We will bound the area of the individual cells from below. To achieve this, we divide the cells into two groups, depending on whether or not they share a point with one of the sides of the square.

  Case 1: The cell does not share a point with a side of the square. Then the area of a regular hexagon drawn around the circle in the cell provides a lower bound for the area of the cell. This follows from the following facts. First, among the polygons with k sides drawn around a circle, the regular polygon has the smallest area. Secondly, out of the equilateral triangle, the square, the regular pentagon and the regular hexagon drawn around a circle, the hexagon has the smallest area. In a special case, the area of the cell equals the area of the respective regular hexagon—namely when a circle of the packing is surrounded by six other circles. From this it follows that for Case 1 a lower bound for the area of the cells is the area of a regular hexagon of side $\overline{r}_n$, that is $2\sqrt{3}\overline{r}_n^2$.

  Case 2: The cell shares a point with one or more sides of the square. Although the lower bound obtained in case 1 is also valid for these types of cells, this bound can be improved because the cell makes contact with at least one side of the square: First, notice that the convex hull of the circle centres of the packing is located in the square $[\overline{r}_n, 1-\overline{r}_n]^2$. Let us denote that region of the unit square by R, which is outside of this inner square, i.e. $R := [0,1]^2 \setminus [\overline{r}_n, 1-\overline{r}_n]^2$. It is easy to see

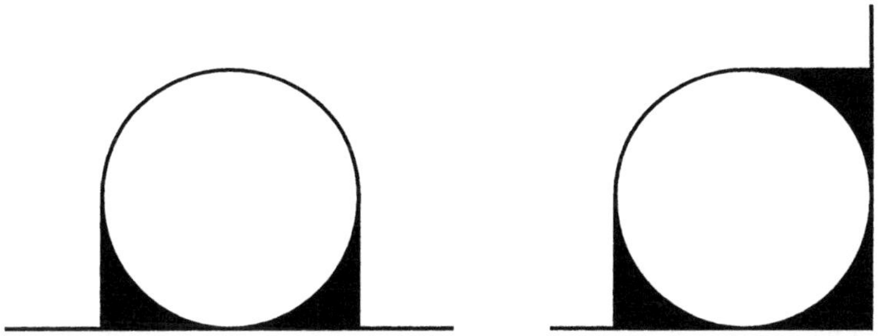

Fig. 3.2. Gaps occurring for circles touching the sides of the square.

that the densest possible way of packing circles into R is to place $4\sqrt{n}-4$ circles on the sides of the unit square.

In this packing, the non-covered regions of R have two different types of 'gaps': the regions defined by two circles and one side of the square, and the regions defined by two sides and a circle located in the corner of $[0,1]^2$ (see Figure 3.2). The sum of the non-covered areas is a lower bound for the area of the non-covered regions of R *for every optimal circle packing* in $[0,1]^2$. A gap increases the estimated lower bound of a regular hexagon by $(2-\sqrt{3})\overline{r}_n^2$. The overall total gap is $(4\lfloor\sqrt{n}\rfloor-2)(2-\sqrt{3})\overline{r}_n^2$. Then the sum of the lower bounds of Dirichlet cells is less than 1 (the area of the unit square), so

$$\overline{r}_n \leq \frac{1}{\sqrt{2\sqrt{3}n+(4\lfloor\sqrt{n}\rfloor-2)(2-\sqrt{3})}}.$$

- $U_2(n)$: Using the Oler Theorem [90], J. H. Folkman and R. L. Graham proved the following statement [29]: if X is a compact convex subset of the plane, then the number of points in X with a mutual distance of at least 1 cannot be more than

$$\frac{2}{\sqrt{3}}A(X)+\frac{1}{2}P(X)+1, \tag{3.2}$$

where $A(X)$ is the area, and $P(X)$ is the perimeter of X, respectively.

Applying this result to an $A'(1,\overline{\sigma}_n)$ associated point arrangement (that is, for $X=[0,\overline{\sigma}_n]^2$, $A(X)=\overline{\sigma}_n^2$, and $P(X)=4\overline{\sigma}_n$), we obtain

$$n \leq \frac{2}{\sqrt{3}}\overline{\sigma}_n^2+2\overline{\sigma}_n+1.$$

From the relation $\overline{\sigma}_n=(1-2\overline{r}_n)/2\overline{r}_n$ of Corollary 2.15, the lower bound

$$\overline{r}_n \leq \frac{1+\sqrt{1+\frac{2}{\sqrt{3}}(n-1)}}{2n+2\sqrt{1+\frac{2}{\sqrt{3}}(n-1)}}$$

follows, i.e. $U_2(n)$ is also a valid upper bound for $\overline{r}_n$. This completes the proof.

□

T. Tarnai and Zs. Gáspár in some special cases improved the previous upper bound in [122].

3.3 Asymptotic estimation and its error

Based on Corollary 2.15 $\left(\overline{m}_n = \frac{2\overline{r}_n}{1-2\overline{r}_n}\right)$, Theorems 3.1 and 3.2 suggest the following inequality bounds:

$$\sqrt{\frac{2}{\sqrt{3}n}} \leq \overline{m}_n \leq \frac{1}{n-1} + \sqrt{\frac{1}{(n-1)^2} + \frac{2}{\sqrt{3}(n-1)}}.$$

If we multiply these inequalities by $\sqrt{n}$, and let n tend to infinity, then

$$\lim_{n\to\infty} \sqrt{n}\overline{m}_n = \sqrt{\frac{2}{\sqrt{3}}}$$

holds, and this means $\overline{m}_n$ is asymptotically equal to $\sqrt{\frac{2}{\sqrt{3}n}}$ [14].

Theorem 3.3. *An upper bound of the absolute error for the* $\overline{m}_n$ $(\approx \sqrt{\frac{2}{\sqrt{3}n}})$ *asymptotic estimation is*

$$\frac{2}{n-1}.$$

PROOF. The above inequality

$$\frac{1+\sqrt{1+\frac{2}{\sqrt{3}}(n-1)}}{n-1} - \sqrt{\frac{2}{\sqrt{3}n}} < \frac{2}{n-1}$$

is equivalent to

$$\sqrt{1+\frac{2}{\sqrt{3}}(n-1)} - \sqrt{\frac{2(n-1)^2}{\sqrt{3}n}} < 1.$$

After a short calculation we obtain the following relation:

$$1 + \frac{2}{\sqrt{3}}(n-1) < 1 + \frac{2(n-1)^2}{\sqrt{3}n} + 2\sqrt{\frac{2(n-1)^2}{\sqrt{3}n}}.$$

This in turn implies that

$$1 < \frac{n-1}{n} + \sqrt{\frac{6}{\sqrt{3}n}}.$$

This is equivalent to $\frac{\sqrt{3}}{6} < n$, but this is obviously true. □

The theoretical lower bounds can be easily improved with the help of computer-aided global optimization methods. In the following chapters these kinds of algorithms will be studied.

4

Approximate Circle Packings Using Optimization Methods

In this chapter we describe a stochastic optimization algorithm for finding good approximate circle packings. With our procedure for instance, we were able to find most of the optimal solutions for each of the problems previously solved and reported in the literature. For $n = 32, 37, 47, 62,$ and 72 the algorithm found better solutions than those reported previously. The arrangements obtained were validated by interval arithmetic computations [11]. First of all let us examine an earlier approach to the problem.

4.1 An earlier approach: energy function minimization

As we saw in Chapter 2, the point arrangement problem expressed a mathematical programming task that can be formalized in the following way:

$$\max \min_{1 \le i < j \le n} \|s_i - s_j\|$$

subject to

$$s_i \in [0,1]^2, \quad 1 \le i \le n.$$

By virtue of $\min_{1 \le i < j \le n} \|s_i - s_j\| = \lim_{m \to -\infty} \left(\sum_{1 \le i < j \le n} \|s_i - s_j\|^m \right)^{\frac{1}{m}}$, the problem is relaxed to

$$\min_{s_i \in [0,1]^2,\ 1 \le i \le n} \sum_{1 \le i < j \le n} \frac{1}{\|s_i - s_j\|^m}.$$

This objective function can be interpreted as a potential or energy function. A physical analogue of this approach is to treat the points as electrical charges (all positive or all negative) which repel each other. If the minimal distance between the charged particles increases, the corresponding value of the potential function decreases. K. J.

Nurmela and P. R. J. Östergård [86] used a similar energy function with large values of m:

$$\sum_{1 \le i < j \le n} \left(\frac{\lambda}{||s_i - s_j||^2} \right)^m .$$

Introducing $x_i = \sin(x_i')$ and $y_i = \sin(y_i')$, this can be transformed into an unconstrained optimization problem with variables x_i', y_i', where the coordinates of the centres of the circles satisfy the constraints $-1 \le x_i \le 1, -1 \le y_i \le 1$.

K. J. Nurmela and P. R. J. Östergård published candidate packings of up to 50 circles using a combination of Goldstein–Armijo backtracking linear search and the Newton method for the optimization [86].

4.2 The TAMSASS-PECS algorithm

The TAMSASS-PECS algorithm is based on the Threshold Accepting method [20] and on our modified version of SASS (Single Agent Stochastic Search) [50, 72, 73, 98, 109] (MSASS). The TAMSASS-PECS algorithm has specifically been designed to handle the 'packing n equal circles in a square' problem, but it can easily be modified to pack equal circles in other regular shapes.

The Threshold Accepting framework and the SASS local search algorithm are used for finding a local maximum of a function defined over a search space. As for the stochastic local search algorithms, the probability of finding a global solution increases with the number of trial points. Here these methods are based on stepwise improvements of the current solution by movements in the solution space. Now the same notation and definitions will be used as in [85].

Definition 4.1. *Let $S_F \subseteq S$ be the set of feasible solutions for a maximization problem and S the search space. For a bounded S the goal of the* maximization problem *is to:*

$$\text{maximize } f(s) \ \text{ subject to } s \in S,$$

where $f : S \to \mathbb{R}$ is the objective (cost) function and $\mathbb{R}$ is the set of real numbers.

Definition 4.2. *A* move *is defined as :*

$$d : S(d) \to S,$$

where $S(d) \subseteq S$ is the domain of the move. The set of the moves of the problem is D. We assert that the union of the domains for each move sequence in D is the solution set, thus there is no point in the search space that cannot be reached from any other points using multiple moves.

We will denote the operation of a move d to a candidate solution by $s \in S$ by $d(s)$.

Algorithm 1 : The Threshold Accepting Algorithm

Select an initial solution s
while (stopping criterion not satisfied)
    while (inner loop criterion not satisfied)
        Select $s' \in N(s)$
        $\Delta = f(s) - f(s')$
        if $\Delta \leq T_h$ **then** $s = s'$
        Decrease threshold T_h
    end while
end while
return the best found solution

Definition 4.3. *A candidate solution s' is a* neighbour *of the candidate solution s, if*

$$s' = d(s) \text{ for some } d \in D.$$

The neighbourhood $N(s)$ of a solution $s \in S(d)$ is the union of all the neighbours of s:

$$N(s) = \bigcup_{d \in D} d(s).$$

Definition 4.4. *A candidate solution s is a* global optimum *if there is no other candidate solution s' in the search space S such that $f(s') > f(s)$. A candidate solution s is a* (weak) local optimum *if there is no other candidate solution s' in the neighbourhood $N(s)$ of s such that $f(s') > f(s)$. In case all the neighbours of s have a value of the objective function smaller than $f(s)$, then s is a* strong local optimum.

Formally, the Threshold Accepting algorithm [20] is very similar to the simulated annealing algorithms. In Threshold Accepting we accept every move that leads to a new solution not much worse than the current solution. To be more specific if s is the current solution, the proposed next solution $s' \in N(s)$ is accepted as the next solution if

$$\Delta = f(s) - f(s') < T_h,$$

where $T_h > 0$ is the threshold level. During the optimization process the threshold level has been gradually decreased like the temperature in simulated annealing. In Threshold Accepting a move which would give a solution much worse than the current solution is rejected, unlike in simulated annealing. The convergence of the Threshold Accepting algorithm has been proved for some cases [2].

The Threshold Accepting method is described in Algorithm 1. The inner loop is performed a constant L times and the stopping criterion is fulfilled when the final value of the threshold level is reached. A new candidate solution can also be taken when Δ is equal to the threshold level, keeping the sideways moving with the same function value allowed. In [20] a more adaptive inner loop stopping criterion is worked with: it ends the inner loop after no improvements were made in the last steps.

Algorithm 2 : The Single Agent Stochastic Search

Proc $SASS(s_0, ex, ct, Scnt, Fcnt, \sigma_{sup}, \sigma_{inf}, NIter)$
var $b_0 = 0; k = 1; scnt = 0; fcnt = 0; \sigma_0 = 1;$

while ($k < NIter$) **do**
  if ($scnt > Scnt$) **then** $\sigma_k = ex \cdot \sigma_{k-1}$
  elsif ($fcnt > Fcnt$) **then** $\sigma_k = ct \cdot \sigma_{k-1}$
    elsif ($\sigma_{k-1} < \sigma_{inf}$) **then** $\sigma_k = \sigma_{sup}$
      else $\sigma_k = \sigma_{k-1}$
      endif
    endif
  endif
  $\xi_k \sim N(b_k, \sigma_k \mathbf{I})$
  if $f(s' = s_k + \xi_k) < f(s_k)$
    then
      $s_{k+1} = s'$
      $b_{k+1} = 0.2b_k + 0.4\xi_k$
      $scnt = scnt + 1$
      $fcnt = 0$
    elsif $f(s' = s_k - \xi_k) < f(s_k)$
      then
        $s_{k+1} = s'$
        $b_{k+1} = b_k - 0.4\xi_k$
        $scnt = scnt + 1$
        $fcnt = 0$
      else
        $s_{k+1} = s_k$
        $b_{k+1} = 0.5b_k$
        $fcnt = fcnt + 1$
        $scnt = 0$
      endif
  endif
  $k = k + 1$
end while

Random optimization is traditionally based on single agent stochastic search strategies [73]. S.S. Rao [98] and D.C. Karnop [50] used a uniform random variable as the move function, $d(s)$. J. Matyas utilized Gaussian perturbations for the move function with a bias term to direct the search, $\xi \sim N(b, \sigma \mathbf{I})$ [72]. F.J. Solis and J.B. Wets [109] enhanced this approach by evaluating the objective function at $s' = s - \xi s$ if the evaluation at $s' = s + \xi s$ does not improve the current value of the objective function and incorporated a variable perturbation variance. The bias term and this additional function evaluation both serve as stochastic equivalents for incorporating momentum and gradient information. The SASS method is shown in Algorithm 2.

The variance of the perturbation size (ξ), is determined by the number of repeated successes and failures, $scnt$ and $fcnt$, respectively, when selecting a neighbour that decreases the value of the objective function. Note that conditions in the third line are mutually exclusive. The contraction (ct) and expansion (ex) constants as well as the upper and lower bounds on the standard deviation of the random perturbation ($\sigma_{sup}, \sigma_{inf}$) are set by the user. The standard deviation of the perturbation is increased when it falls below a given lower bound σ_{inf} (third line). The contraction and expansion values of σ are set when the number of successes ($scnt$) or failures ($fcnt$) are greater than the supplied user constants, $Scnt$ and $Fcnt$, respectively.

The values 0.4 and 0.2, which affect the bias value were recommended in F.J. Solis and J.B. Wets's paper [109]. Other values used are: $ex = 2$, $ct = 0.5$, $Scnt = 5$, $Fcnt = 3$, $\sigma_{sup} = 1.0$ and $\sigma_{inf} = 10^{-5}$. The standard deviation σ specifies the size of the sphere that most likely contains the perturbation vector and the bias term b locates the centre of the sphere based on directions of a past success. In line 16 of Algorithm 2 a reverse strategy, seeking a better solution in the opposite direction of the originally failed move, is performed. The stopping criterion is based on the number of iterations done.

The TAMSASS-PECS algorithm (Algorithm 3) is based on a Threshold Accepting method and on a modified version of SASS (MSASS), which was especially designed to improve the current solution s of the 'packing equal circles in a square' problem. The objective function returns the minimum distance from the centre of a circle to the other centres, and the problem in question consists of maximizing this minimum distance for all centres. Thus the program does a maximization of the minimum distance between circle centres in an iterative way.

MSASS, which controls the location of a point s_i, was constructed using the parameters provided by TAMSASS-PECS. This perturbation approach is intended to increase the minimum value of the distances ($d_{i,j}$) between s_i and any point s_j ($1 \leq j \neq i \leq n$). It moves the point s_i to a new location s_i' and computes the value of the new minimum distance $d_{i,j}'$. Following the Threshold Accepting strategy, a move is accepted if $d_{i,j} - d_{i,j}' < d_{i,j}T_h$, where $d_{i,j}T_h$ is the threshold level and $T_h > 0$. New trial locations of points as s_i' are restricted to the neighbourhood of the current location of the point s_i. This neighbourhood has a normal distribution $N(s_i, \sigma\mathbf{I})$ around s_i. Until the Threshold Accepting condition is met, new locations for the point s_i should be tested using the classical SASS algorithm, although the number of trials is limited by a set maximum value (see Algorithm 4).

The TAMSASS-PECS algorithm starts with a pseudorandom initial solution (the location of the n points s_i, $1 \leq i \leq n$, which are generated by dividing the square into $t = k \times k$ non-overlapping tiles, where $k = \lceil \sqrt{n} \rceil$, and $\lceil . \rceil$ denotes the smallest integer not less than the argument. Notice that the number of points n is less than or equal to the number of tiles t. The first point is located randomly at the centre of the first or second tile. The following points are located at the centre of a tile which is separated from the previous tile by one free tile in consecutive order. The remaining points are randomly allocated to the free tiles (at most one point in one tile). However, based on our experience if the threshold level is not too small, then the initial solution is not crucial because large random movements are also allowed.

Algorithm 3 : TAMSASS-PECS. Threshold accepting algorithm adapted to the packing of equal circles in a square

Select an initial solution s
Select an initial threshold T_h
Select an initial standard deviation σ
while ($\sigma > \sigma_{final}$) **do**
  while (All centres are not visited) **do**
    $s = MSASS(s, \sigma, T_h, NextCentre(s))$
  end while
  Decrease threshold T_h
  Decrease standard deviation σ
end while
return the best found solution

Algorithm 4 : MSASS. Modified Single Agent Stochastic Search

Proc $MSASS(s, \sigma_0, T_h, i)$
var $scnt = 0$; $fcnt = 0$; $Fcnt = 3$, $ct = 0.5$; $k = 1$;

while ($fcnt < 4 \cdot Fcnt$ **and** $scnt = 0$) **do**
  if ($fcnt > Fcnt$) **then** $\sigma_k = ct \cdot \sigma_{k-1}$
  else $\sigma_k = \sigma_{k-1}$
  generate a random number ξ_k with $N(0, \sigma_k \mathbf{I})$ distribution
  $s'(i) = s_k(i) + \xi_k$
  if ($f(s') \geq f(s_k(i)) \cdot T_h$)
    then
      $s_{k+1}(i) = s'(i)$
      $scnt = scnt + 1$
    else
      $s'(i) = s_k(i) - \xi_k$
      if ($f(s') \geq f(s_k(i)) \cdot T_h$)
        then
          $s_{k+1}(i) = s'(i)$
          $scnt = scnt + 1$
        else
          $s_{k+1}(i) = s_k(i)$
          $fcnt = fcnt + 1$
      endif
  endif
$k = k + 1$
end while
return s

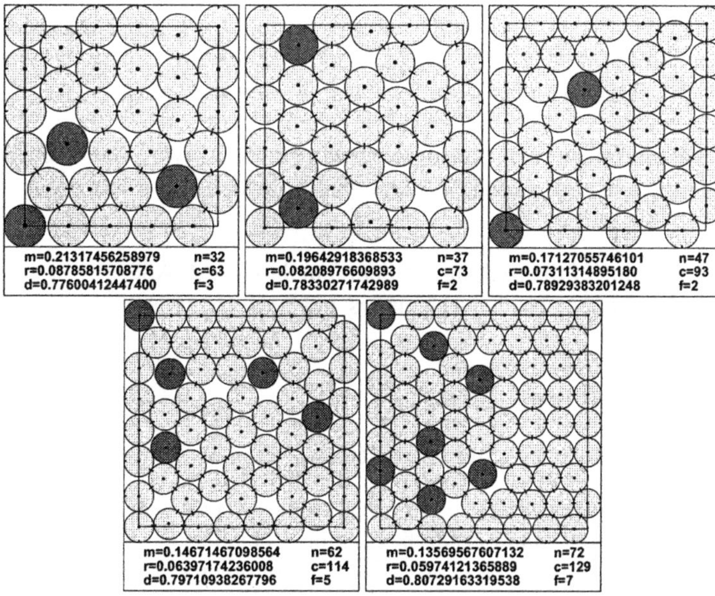

Fig. 4.1. Some improved packings for $n = 32, 37, 47, 62$, and 72 circles found by the TAMSASS-PECS method.

The initial value for the threshold level is $T_h = 0.02$, and the standard deviation σ is equal to the common diameter of the tiles. The TAMSASS-PECS algorithm tries to improve the current solution via an iterative procedure. At each step the MSASS subroutine is executed for all the n points using the same values for both the standard deviation σ and threshold level T_h. The criterion for halting this iterative procedure is based on the value of the standard deviation, which is decreased by a factor of 0.99 at each iteration. The threshold level is also decreased by the same factor.

At each iterative step, the MSASS algorithm is run for all the n points in an increasing order, which is determined by the value of the minimum distance from a point to the remaining points ($d_{i,j}$). The value of the minimum distance is updated after each execution of MSASS. The description of the TAMSASS-PECS method is included in Algorithms 3 and 4.

4.3 Computational results

In this section results obtained from using the TAMSASS-PECS algorithm will be compared with the best results found and previously reported in the literature (see Figure 4.1). The programs were coded in C and run under Linux on a PC with Pentium II (266 Mhz) processor. Since the problems are very hard, sometimes the solution of a problem required multiple runs, thus the CPU times ranged from a few seconds to a few hours. Details of the numerical results and graphical representations are available at

`http://www.inf.u-szeged.hu/~pszabo/Packing_circles.html`

Tables 4.1-4.3 show some results of the problem for $2 \leq n \leq 100$. In the columns we refer to the numerical results of our algorithm with headers m_n, c_n, f_n and d_n. Here:

m_n: is the solution of the packing problem found by TAMSASS-PECS,
c_n: the number of contacts between circles and between circles and the sides of the square,
f_n: the number of free circles, and
d_n: the density value of the packing.

Now let us summarize the results of Tables 4.1 to 4.3:

- In 20 out of 59 cases we found the same results as reported in the literature. Most of them are optimal solutions of the packing problem.
- In 34 cases our results were worse than those of other authors. Nevertheless, for several cases our algorithm obtained similar final geometrical arrangements, and the difference was only in the accuracy of the calculated numerical values.
- For $n = 32, 37, 47, 62$, and 72 our results improved all of those previously published.
- 40 new results were generated. In each case here the value of m_n was found to be better than or equal to the new lower bound described in [118].

The accuracy of our results is 10^{-14}, and the feasibility of these solutions was verified by using interval arithmetic. According to these tests, then, there are no better packings in the very close neighbourhood of the given results. But it should be noted that these interval verification tests cannot prove global optimality.

Table 4.1. The best known results for packing circles in a unit square with $2 \leq n \leq 35$ using the TAMSASS-PECS algorithm.

n	m_n	c_n	f_n	d_n
2	1.41421356237309	5	0	0.53901208445264
3	1.03527618041008	7	0	0.60964480874135
4	1.00000000000000	12	0	0.78539816339744
5	0.70710678118654	12	0	0.67376510556580
6	0.60092521257733	13	0	0.66395690946413
7	0.53589838486224	14	1	0.66931082684079
8	0.51763809020504	20	0	0.73096382525390
9	0.50000000000000	24	0	0.78539816339744
10	0.42127954398390	21	0	0.69003578526417
11	0.39820731023684	20	2	0.70074157775610
12	0.38873012632301	25	0	0.73846822388404
13	0.36609600769623	25	1	0.73326469490355
14	0.34891526037401	32	1	0.73567925554268
15	0.34108137740210	36	0	0.76205601092668
16	0.33333333333333	40	0	0.78539816339744
17	0.30615398530033	34	1	0.73355026330232
18	0.30046260628866	38	0	0.75465335787566
19	0.28954199199356	37	2	0.75230789673638
20	0.28661165235168	44	0	0.77949368686760
21	0.27181168718966	40	2	0.75335522651101
22	0.26795835833157	42	1	0.77167991577529
23	0.25881904510252	56	0	0.76363103212612
24	0.25433309503024	56	0	0.77496325975782
25	0.25000000000000	60	0	0.78539816339744
26	0.23873475724121	56	2	0.75846909048393
27	0.23584952830050	55	0	0.77231145646250
28	0.23053540627071	56	1	0.77185363595323
29	0.22688290074420	65	1	0.77890624177970
30	0.22450296453108	65	0	0.79201902646073
31	0.21754726920445	54	4	0.77729734717346
32	0.21317456258979	63	3	0.77600412447400
33	0.21132833175032	64	1	0.78885198102597
34	0.20560464675956	80	0	0.77664906433227
35	0.20276360086322	80	0	0.78122721299871

Table 4.2. The best known results for packing circles in a unit square with $36 \leq n \leq 70$ using the TAMSASS-PECS algorithm.

n	m_n	c_n	f_n	d_n
36	0.20000000000000	84	0	0.78539816339744
37	0.19642918368533	73	2	0.78330271742989
38	0.19534202418353	74	0	0.79704064569288
39	0.19436506314440	80	0	0.81117902717206
40	0.18817551865770	85	2	0.78797949519086
41	0.18609948922999	97	1	0.79272373684650
42	0.18427707211709	90	0	0.79868427865342
43	0.18019112440214	83	1	0.78726408457444
44	0.17863916198187	79	4	0.79384217677293
45	0.17571631417546	94	3	0.78944426849399
46	0.17445933585066	90	1	0.79718693728810
47	0.17127055746101	93	2	0.78929383201248
48	0.16938050793544	97	1	0.79094499029400
49	0.16738607686832	120	1	0.79121698952690
50	0.16645461022968	102	0	0.79967929970534
51	0.16561390218764	90	2	0.80861948195609
52	0.16538621483322	103	0	0.82253064459417
53	0.16264607633837	93	4	0.81462525898420
54	0.15913951414578	109	2	0.79940760435030
55	0.15755573439859	105	2	0.80027118200754
56	0.15615650046214	119	0	0.80235172950297
57	0.15474723074928	107	3	0.80396408840952
58	0.15130206629471	109	6	0.78673600284850
59	0.15029765718520	118	3	0.79108990822535
60	0.14945461308474	125	3	0.79666571193460
61	0.14854408720568	119	1	0.80137375361858
62	0.14671467098564	114	5	0.79710938267796
63	0.14667828199655	118	2	0.80961563357266
64	0.14532088220301	132	3	0.80922920583758
65	0.14456113665810	112	5	0.81438235586667
66	0.14380319561470	120	2	0.81934777745264
67	0.14304397643538	118	5	0.82409628091674
68	0.14288350731175	115	6	0.83475507341762
69	0.13993726983993	117	3	0.81666484211077
70	0.13787246738894	114	5	0.80715295095497

Table 4.3. The best known results for packing circles in a unit square with $71 \leq n \leq 100$ using the TAMSASS-PECS algorithm.

n	m_n	c_n	f_n	d_n
71	0.13593878298879	133	3	0.79859232191938
72	0.13569567607132	129	7	0.80729163319538
73	0.13389894938934	135	5	0.79949982777920
74	0.13265445602583	141	6	0.79720572964583
75	0.13241101073542	108	11	0.80536208356566
76	0.13094762506744	127	9	0.80022799448052
77	0.13047875975659	120	6	0.80562963362431
78	0.13045661660359	157	3	0.81584734690826
79	0.12976111574313	131	8	0.81852673455691
80	0.12886897358290	147	5	0.81882205826177
81	0.12833078028621	170	3	0.82293152331831
82	0.12646140351791	145	4	0.81168420885740
83	0.12609559154647	167	3	0.81736730901968
84	0.12555071002762	152	5	0.82087566500715
85	0.12511540460076	146	4	0.82553641420847
86	0.12500381019147	123	7	0.83392471939050
87	0.12230707667418	158	7	0.81150070836153
88	0.12099893725078	162	10	0.80523985787757
89	0.12080111590653	140	14	0.81201615746621
90	0.11988928132785	160	7	0.81010801424488
91	0.11916044953221	172	8	0.81023468623028
92	0.11854559378259	175	4	0.81159835604241
93	0.11827868195489	167	6	0.81711971813354
94	0.11733029717908	158	10	0.81409469757729
95	0.11668436071663	174	4	0.81466286707370
96	0.11633897630581	192	5	0.81887839722130
97	0.11618529717218	179	4	0.82545113090290
98	0.11610339843903	170	7	0.83290785939723
99	0.11601228667056	193	2	0.84022403037368
100	0.11456309641300	184	6	0.82979353948491

5

Other Methods for Finding Approximate Circle Packings

In this chapter we give a short overview of some other methods used to find approximate circle packings in the unit square. Several deterministic and stochastic strategies were employed here.

5.1 A deterministic approach based on LP-relaxation

The 'packing equal circles in a square' (PECS) problem can be regarded as an all-quadratic optimization problem, i.e. an optimization problem with constraints that are not necessarily convex quadratic. The hardness of the problem is actually due to the large number of constraints. This approach leads to a rectangular subdivision branch-and-bound algorithm. To provide an upper bound at each node of the branch-and-bound tree, M. Locatelli and U. Raber used the special structure of the constraints and gave an LP-relaxation [60]. They found candidate packings for up to 39 circles that confirm the theoretical optimality within a given accuracy, except when $n = 36$ and 37. Moreover, a new solution for $n = 37$ was found, but not yet proved to be optimal within the given tolerance. The authors also proved some interesting properties of the optimal arrangements.

Theorem 5.1. *There always exists some optimal solution of the point arrangement problem such that at each vertex of the unit square, one and only one of the following statements holds:*

- *a point of the optimal solution is in the vertex, or*
- *two points of the optimal solution belong to the edges specifying the vertex and have distances equal to the optimal one ($\overline{m}_n$).*

The second important property proven is related to the structure of optimal solutions.

Theorem 5.2. *There always exists some optimal solution of the point arrangement problem such that along each edge of the unit square there is no portion of the edge of* *width greater than or equal to twice the optimal distance $\overline{m}_n$ which does not contain any point of the optimal solution.*

B. Addis, M. Locatelli, and F. Schoen implemented a version of the Monotonic Basin Hopping (MBH) stochastic method for the local search SNOPT 6.2, and found many improved circle packings for $n = 88, 106, 108, 115, 116, 130, 133, 134, 135, 146, 155$, and 157 circles [1].

5.2 A perturbation method

D. W. Boll and his coworkers used a stochastic algorithm [5] that gave improved packings for $n = 32, 37, 48$, and 50. A brief outline of their method is:

Step 1: consider n random points in the unit square and define the perturbation size $s = 0.25$,

Step 2: for each point

- *a)* perturb the place of the centre by s in the North, South, East, or West direction,
- *b)* if during this process the distance between the point and its nearest neighbour becomes greater, update the new location of the point,

Step 3: repeat *Step 2* while movable points exist,

Step 4: $s := s/1.5$, if $s > 10^{-10}$, and continue with *Step 2*.

Using this simple algorithm, good candidate packings can be found after some millions of iterations. The authors of the paper [5] found earlier unpublished approximate packings for up to $n = 200$.

5.3 Billiard simulation

To understand this algorithm, let us consider a random arrangement of the points. Draw equal circles around the points without any overlapping. Each circle can be considered as a ball with an initial radius, direction of motion direction, and speed. Start the "balls" (better disks, regarding the two-dimensional nature of our problem) and slowly increase their common radius. The oscillation of each ball during the process will gradually become less and less. The algorithm will stop when the packing or a substructure of the packing becomes fixed. Using a billiard simulation, R. L. Graham and B. D. Lubachevsky [36] reported several new approximate packings for up to 50 circles and for some number of circles beyond.

5.4 Pulsating Disk Shaking (PDS) algorithm

The PDS (Pulsating Disk Shaking) algorithm is similar to the Billiard simulation explained in the previous section. Like the above method it uses a set of congruent disks, therefore it is more geometrically motivated from the "actors" point of view. In each stage the disks are either separated from each other or are in contact (to be more precise the latter means they are just in contact or they overlap). Only distances are considered here. On the other hand, it is well known that good packings were achieved in 3D by real physical shaking of balls. Thus the PDS algorithm itself is based on a more physically motivated "shake-and-rattle" concept.

To explain the method in detail, consider n small equal disks placed in the square such that they do not overlap. Now blow them up with a constant rate until they come into contact between themselves or with the sides of the square. From this time, measure and accumulate the displacements for each pair of disks that would make them just touching (i.e. non-overlapping). Note that this is a pure static and geometric approach, no velocities or moments play a role. One cycle consists here of all $\binom{n}{2}$ pairwise comparisons between two disks and n comparisons with the sides of the square.

For each of these comparisons the displacements are calculated as vectors to separate the partners exactly. All the displacements are accumulated until one cycle is over. Only then will they be applied to the disks which won't yield to a non-overlapping packing in general. So the packing is permanently shaken. As a measure of the progress the total overlap

$$ovl_{\mathrm{t}} = \sum_{ovl \leq 0} ovl$$

is calculated at the end of each cycle where

$$ovl = \begin{cases} d - 2r & \text{for disk–disk contacts,} \\ d - r & \text{for disk–side contacts} \end{cases}$$

(d is the distance between the disk centres or disk centre to side, r is the radius of disks). Clearly for "real" contacts we have $ovl \leq 0$.

If after some cycles no non-overlapping packing can be achieved, the common radius of the disks will be reduced by a constant factor, otherwise it will be increased by a different constant factor. Hence the current radius of the packing is pulsating during the process. The algorithm stops either when the amplitude of the pulsation is small enough, normally equal to the wanted precision of the coordinates, or if the number of cycles exceeds a maximum number. In this way the packing under consideration converges automatically to a more or less rigid one.

A significant reduction of the computational cost can be achieved if contact pairs (disk-to-disk or disk-to-side) are classified as NEAR or FAR. Initially, all pairs are set to NEAR which implies that only such pairs of disks will be checked during the calculation. When after thousands of collisions the mutual distance of a pair is large enough, then the pair is set to FAR which means that they probably won't touch each other in later cycles. While the program is running, the computational cost will become less and less.

It is useful to consider not just random arrangements for the initial packing but also a hexagonal or other regular lattice packing. Sometimes the relationship between the number of circles and the structure of the packing can suggest a good initial configuration, see the finite pattern classes in Chapter 10. Please refer to Section 12.2 for a detailed description of the corresponding C++ language program `csq.c`.

5.5 Computational results

In this section computational results of the packings we found are reported for up to $n = 200$ circles using the PDS algorithm outlined above. These packings were of course found via numerical computations. An important part of the investigations was to provide a guarantee that the arrangements with the given structures really do (and also in a mathematical sense) exist.

A possible way of proving this is based on a system of nonlinear equations which corresponds to the packings. The solution of the system of equations (when it exists) rigorously proves the existence of the packing. To solve the system of nonlinear equations is easy if we can guess the solution. Sometimes the structures of the packings help us to guess the exact coordinates of the circles (e.g. for grid packings). Here any kind of reliable tools can be used (e.g. the software package INTBIS) to solve the system of equations.

Tables 5.1 – 5.3 summarize the current state of affairs. The improved packings compared to other previous results are depicted in bold face, and italics means that there are better known values, but the precise coordinates have not yet been published. In 66 cases we have found similar values to the packings reported in the literature. In 24 cases the results were worse than David Boll et al.'s packings. The number of improved lower bounds for the packings is 110, which shows that our algorithm is a good one.

Table 5.1. Numerical results for $n = 2 - 36$ circles.

n	exact r_n	exact m_n	approx. m_n	d_n
2	$\frac{1}{2}(2-\sqrt{2})$	$\sqrt{2}$	1.4142135624	0.5390120845
3	$\frac{1}{2}(8-5\sqrt{2}+4\sqrt{3}-3\sqrt{6})$	$\sqrt{6}-\sqrt{2}$	1.0352761804	0.6096448087
4	$\frac{1}{4}$	1	1.0000000000	0.7853981634
5	$\frac{1}{2}(-1+\sqrt{2})$	$\frac{1}{2}m_2$	0.7071067812	0.6737651056
6	$\frac{1}{46}(-13+6\sqrt{13})$	$\frac{1}{6}\sqrt{13}$	0.6009252126	0.6639569095
7	$\frac{1}{13}(4-\sqrt{3})$	$4-2\sqrt{3}$	0.5358983849	0.6693108268
8	$\frac{1}{4}(1+\sqrt{2}-\sqrt{3})$	$\frac{1}{2}m_3$	0.5176380902	0.7309638253
9	$\frac{1}{6}$	$\frac{1}{2}$	0.5000000000	0.7853981634
10	–	–	0.4212795440	0.6900357853
11	(see below)	(see below)	0.3982073102	0.7007415778
12	$\frac{1}{382}(-34+15\sqrt{34})$	$\frac{1}{15}\sqrt{34}$	0.3887301263	0.7384682239
13	–	–	0.3660960077	0.7332646949
14	$\frac{1}{33}(6-\sqrt{3})$	$\frac{2}{13}(4-\sqrt{3})$	0.3489152604	0.7356792555
15	$\frac{1}{2}r_3$	$2r_8$	0.3410813774	0.7620560109
16	$\frac{1}{8}$	$\frac{1}{3}$	0.3333333333	0.7853981634
17	–	–	0.3061539853	0.7335502633
18	$\frac{1}{262}(-13+12\sqrt{13})$	$\frac{1}{2}m_6$	0.3004626063	0.7546533579
19	–	–	0.2895419920	0.7523078967
20	$\frac{1}{482}(65-8\sqrt{2})$	$\frac{1}{16}(6-\sqrt{2})$	0.2866116524	0.7794936869
21	–	–	0.2718122554	0.7533577029
22	–	–	0.2679584016	0.7716801121
23	$\frac{1}{2}(-7-5\sqrt{2}+4\sqrt{3}+3\sqrt{6})$	$\frac{1}{4}m_3$	0.2588190451	0.7636310321
24	$\frac{1}{92}(21-5\sqrt{2}+3\sqrt{3}-4\sqrt{6})$	r_3	0.2543330950	0.7749632598
25	$\frac{1}{10}$	$\frac{1}{4}$	0.2500000000	0.7853981634
26	–	–	0.2387347572	0.7584690905
27	$\frac{1}{3022}(-89+40\sqrt{89})$	$\frac{1}{40}\sqrt{89}$	0.2358495283	0.7723114565
28	–	–	0.2305354936	0.7718541114
29	–	–	0.2268829007	0.7789062418
30	$\frac{1}{1202}(126-5\sqrt{10})$	$\frac{1}{75}(20-\sqrt{10})$	0.2245029645	0.7920190265
31	–	–	0.2175472916	0.7772974787
32	–	–	0.2131341934	0.7757618736
33	–	–	0.2113283841	0.7888523039
34	$\frac{1}{2}m_8$	$2r_{23}$	0.2056046467	0.7766490643
35	(see below)	$2r_{24}$	0.2027636009	0.7812272130
36	$\frac{1}{12}$	$\frac{1}{5}$	0.2000000000	0.7853981634

$$r_{11} = \frac{1}{568}\quad 176 - 9\sqrt{2} - 14\sqrt{3} - 13\sqrt{6} - 2\sqrt{-16523 + 12545\sqrt{2} - 9919\sqrt{3} + 6587\sqrt{6}}$$

$$m_{11} = \frac{1}{4}\quad -4 - 3\sqrt{2} + 2\sqrt{3} + 3\sqrt{6} + \quad 4 + \sqrt{2} - 2\sqrt{3} - \sqrt{6}\quad \sqrt{1 + 2\sqrt{2}}$$

$$r_{35} = \frac{1}{772}(112 - 17\sqrt{2} + 8\sqrt{3} - 15\sqrt{6})$$

Table 5.2. Numerical results for $n = 37 - 124$ circles.

n	approx. m_n	d_n	n	approx. m_n	d_n
37	0.1964291842	0.7833027206	**81**	0.1283368559	0.8230005836
38	0.1953423041	0.7970425568	**82**	0.1274269118	0.8227146966
39	0.1943650631	0.8111790272	**83**	0.1264543531	0.8215014743
40	0.1881755220	0.7879795188	**84**	0.1257627022	0.8233399269
41	0.1860995118	0.7927238993	85	0.1253084290	0.8278015474
42	0.1842770721	0.7986842786	*86*	0.1250425340	0.8343840262
43	0.1801911354	0.7872641664	**87**	0.1228265796	0.8176519577
44	0.1786392456	0.7938428078	**88**	0.1220983545	0.8183334905
45	0.1757163141	0.7894442684	**89**	0.1209431296	0.8137202694
46	0.1744593608	0.7971871319	90	0.1204480494	0.8168616061
47	0.1712705638	0.7892938820	**91**	0.1198126022	0.8181738104
48	0.1694054293	0.7911440338	**92**	0.1193622666	0.8216190446
49	0.1673860768	0.7912169895	**93**	0.1183841706	0.8184234765
50	0.1665265773	0.8002721839	**94**	0.1180563847	0.8231316147
51	0.1656183743	0.8086569481	**95**	0.1171194312	0.8201101448
52	0.1653862379	0.8225308420	96	0.1167579991	0.8241689679
53	0.1626480663	0.8146424042	**97**	0.1163574839	0.8276442131
54	0.1591395163	0.7994076230	**98**	0.1165351364	0.8335521941
55	0.1575557475	0.8002712972	**99**	0.1160181348	0.8402999370
56	0.1561565004	0.8023517295	**100**	0.1145801945	0.8300157794
57	0.1547474069	0.8039656738	**101**	0.1137678096	0.8276766917
58	0.1526925313	0.7993307243	**102**	0.1130265581	0.8261170420
59	0.1515619183	0.8026893318	**103**	0.1124324277	0.8263480419
60	0.1495056540	0.7971391564	**104**	0.1119552283	0.8280134249
61	0.1485441266	0.8013741245	**105**	0.1117113035	0.8327015567
62	0.1474526798	0.8041134778	**106**	0.1111479979	0.8330195984
63	0.1468193136	0, 8109737806	*107*	0.1098006344	0.8226089313
64	0.1453677544	0.8096850385	108	0.1091766948	0.8218111889
65	0.1446990147	0.8157400181	109	0.1087836926	0.8240438322
66	0.1438039660	0.8193554530	110	0.1081056069	0.8222742690
67	0.1430855758	0.8245156558	**111**	0.1077672183	0.8250669455
68	0.1429094775	0.8350205982	**112**	0.1072831265	0.8257591285
69	0.1399481818	0.8167765737	*113*	0.1068253164	0.8267201679
70	0.1379067766	0.8075060206	**114**	0.1066053963	0.8309359404
71	0.1366129972	0.8055769545	**115**	0.1059840023	0.8294126663
72	0.1358499279	0.8089083021	**116**	0.1056270416	0.8315355075
73	0.1347098276	0.8080563922	**117**	0.1053200675	0.8342993237
74	0.1339986726	0.8115167751	**118**	0.1051451721	0.8389032937
75	0.1324888136	0.8061980168	**119**	0.1050811970	0.8450812908
76	0.1317300376	0.8086999566	**120**	0.1050454468	0.8516581650
77	0.1308410780	0.8095910198	**121**	0.1034896517	0.8358581333
78	0.1304607726	0.8158933303	*122*	0.1028022705	0.8328448925
79	0.1299652027	0.8208069227	**123**	0.1021525019	0.8298691626
80	0.1296133854	0.8272178885	**124**	0.1015480773	0.8276525614

Table 5.3. Numerical results for $n = 125 - 200$ circles.

n	approx. m_n	d_n	n	approx. m_n	d_n
125	0.1012318916	0.8296158463	**163**	0.0885685524	0.8474713225
126	0.1009062359	0.8313727312	**164**	0.0881985853	0.8461369376
127	0.1005031141	0.8318980160	**165**	0.0878493719	0.8451107320
128	0.1000307339	0.8312987414	**166**	0.0875180593	0.8443458471
129	0.0993263118	0.8270941821	**167**	0.0873073510	0.8456746787
130	0.0988937215	0.8269119921	**168**	0.0871824504	0.8485011578
131	0.0986288136	0.8292143817	**169**	0.0863882436	0.8392971739
132	0.0982888835	0.8303084177	**170**	0.0859789843	0.8369135174
133	0.0979321354	0.8310764541	*171*	0.0855822297	0.8346948524
134	0.0978107863	0.8354360146	*172*	0.0853489718	0.8353646965
135	0.0975218520	0.8371459461	**173**	0.0851552703	0.8367106190
136	0.0973414248	0.8405056543	**174**	0.0850212754	0.8391079841
137	0.0971962814	0.8443861367	**175**	0.0847072864	0.8381936262
138	0.0965684977	0.8405593584	**176**	0.0845723809	0.8405094143
139	0.0963607140	0.8433304280	**177**	0.0842258678	0.8389085248
140	0.0960210816	0.8439433382	**178**	0.0839212683	0.8380279161
141	0.0956706024	0.8443178921	**179**	0.0838518837	0.8414507200
142	0.0954741493	0.8471212060	**180**	0.0837056682	0.8434307533
143	0.0953634647	0.8512820167	**181**	0.0833661059	0.8417768920
144	0.0940769937	0.8362256095	*182*	0.0831953206	0.8432289693
145	0.0937839766	0.8372440265	*183*	0.0829799252	0.8438130489
146	0.0932472250	0.8342146320	*184*	0.0828621097	0.8462006652
147	0.0927445712	0.8316620732	*185*	0.0828104531	0.8498202051
148	0.0924063478	0.8317384492	**186**	0.0827835097	0.8539004243
149	0.0923037281	0.8356565148	**187**	0.0827219524	0.8573124997
150	0.0920498844	0.8370331951	**188**	0.0826931189	0.8613421953
151	0.0916459689	0.8358530518	**189**	0.0819385675	0.8513795535
152	0.0911461597	0.8329987760	**190**	0.0815394198	0.8481916850
153	0.0908262169	0.8330913575	**191**	0.0811842378	0.8457992069
154	0.0906340269	0.8352857436	**192**	0.0809200493	0.8451158427
155	0.0903257963	0.8354733542	*193*	0.0805888840	0.8430949394
156	0.0899733630	0.8348541653	*194*	0.0804752001	0.8452518615
157	0.0898047212	0.8373181432	**195**	0.0804115030	0.8483644354
158	0.0896806181	0.8405154516	**196**	0.0800186870	0.8450185742
159	0.0895956726	0.8443652173	*197*	0.0795255746	0.8396607811
160	0.0894304663	0.8468018971	*198*	0.0792782634	0.8390666657
161	0.0893487418	0.8506653983	*199*	0.0790085605	0.8379950934
162	0.0888450720	0.8471092098	**200**	0.0789188489	0.8404343621

6

Interval Methods for Validating Optimal Solutions

In the next few chapters we shall discuss special types of numerical methods for *finding optimal packing configurations in a reliable way.* Here 'reliability' means that although the optimality statements are made based on computer runs, the results are still valid and rigorous from a mathematical point of view. Hence these methods will be considered as computer-assisted proofs.

To ensure mathematical correctness for the implementation and execution of computer algorithms, our numerical methods are based exclusively on *interval analysis.* The central theoretical property of interval calculations is that they allow one to determine reliable (sometimes called guaranteed) enclosures for the range of a function over a given domain. With a proper implementation, this central property remains valid even in practical computations, even in the presence of numerical (representation and rounding) errors occurring in binary floating-point computer arithmetic.

After describing the most efficiently applicable problem formulation for the numerical methods, we will give an introduction to the basics of interval analysis, and sketch an interval global optimization algorithm framework based on the 'branch–and–bound' (also known as 'divide and conquer') principle.

The technical details and implementation of the problem-specific optimization algorithms are presented in Chapters 7 to 9. In Chapter 7, our first solely interval-based method is outlined, then in Chapters 8 and 9 the details of a more advanced method are presented. This latter method finally led to the *complete numerical and geometrical (structural) solution of the previously unsolved problem instances of packing 28, 29, and 30 circles into a square* (cf. Table 1.2 in Section 1.2.8).

6.1 Problem formulation for the interval algorithms

Let us start from the classical description of the point arrangement problem (see Problem 2.2): *place n points in a unit square, such that the minimum pairwise distance between the points is maximal.* In formal terms

Problem 6.1.

$$\text{Maximize} \min_{1\leq i\neq j\leq n} \sqrt{(x_i - x_j)^2 + (y_i - y_j)^2}, \tag{6.1}$$

$$\text{subject to} \quad 0 \leq x_i, y_i \leq 1, \quad i = 1, 2, \ldots, n,$$

where the unit square is $[0, 1]^2$, and the ith point is located at (x_i, y_i).

The problem was formulated this way for several reasons. One is that since the computer representation of points is easier than that of circles, we prefer a point packing variant to the classical circle packing problem. Another is that Problem 6.1 leads to a bound-constrained optimization problem (where the only constraints are the bounds on the variables). This can be handled much easier than the circle packing problem itself or the associated circle and point arrangement variants of Chapter 2, all of which result in more general equality and/or inequality constrained problems.

On the other hand, during the numerical computations it is more appropriate to deal with squared distances rather than simple Euclidean distances. This saves the square root operation during the function evaluations, so we compute the distance values only when they are explicitly required—e.g. in procedures which utilize the geometric properties of the point arrangement problem class. Thus for the numerical algorithms of the next chapters we shall consider the point packing problem with *squared distances*:

Problem 6.2.

$$\text{Maximize} \min_{1\leq i\neq j\leq n} (x_i - x_j)^2 + (y_i - y_j)^2, \tag{6.2}$$

$$\text{subject to} \quad 0 \leq x_i, y_i \leq 1, \quad i = 1, 2, \ldots, n,$$

where the unit square is $[0, 1]^2$, and the ith point is located at (x_i, y_i).

To simplify the notation, the *pairwise squared distances* of the point packing problem will be denoted by d_{ij}, that is, $d_{ij} = (x_i - x_j)^2 + (y_i - y_j)^2$, $1 \leq i \neq j \leq n$. In general, the *squared Euclidean distance between two points* $x, y \in \mathbb{R}^n$ *will be* denoted by $d(x, y)$.

Since Problem 6.2 is derived from Problem 6.1 by applying a strictly monotonic transformation on the objective function, the optimizers of the two problems are the same. Thus, from Theorem 2.9 it also follows that Problem 6.2 is equivalent to the traditional circle packing problem.

Consequently, from now on we will consider the objective function

$$f_n : \mathbb{R}^{2n} \to \mathbb{R}, \quad f_n(x, y) = \min_{1\leq i\neq j\leq n} (x_i - x_j)^2 + (y_i - y_j)^2, \tag{6.3}$$

to be maximized over $0 \leq x_i, y_i \leq 1$, $i = 1, 2, \ldots, n$.

6.2 Interval analysis

In this section we give a brief outline on interval analysis that is used as our basic tool to produce mathematically valid statements on computers. More information on the foundations of interval computations can be found in textbooks like [44, 81, 83, 84, 99, 100].

The set of real-type, compact *intervals* will be denoted by $\mathbb{I}$, where for all $A \in \mathbb{I}$, intervals $A = [\mathrm{lb}(A), \mathrm{ub}(A)] = \{a \in \mathbb{R} \mid \mathrm{lb}(A) \le a \le \mathrm{ub}(A)\}$. Here $\mathrm{lb}(A) \in \mathbb{R}$, and $\mathrm{ub}(A) \in \mathbb{R}$ denote the *lower* and *upper bounds* of A, respectively. If $\mathrm{lb}(A) = \mathrm{ub}(A)$, then we call A a *point interval*. For a given set of reals $S \subseteq \mathbb{R}$, let $\mathbb{I}(S)$ denote the set of all intervals in S. The *width* of an interval is defined by $w(A) := \mathrm{ub}(A) - \mathrm{lb}(A)$.

The real *arithmetic operations* can be extended to intervals by applying the general definition $A \circ B := \{a \circ b \mid a \in A,\ b \in B\}$. In particular,

$$\begin{aligned}
A + B &= [\mathrm{lb}(A) + \mathrm{lb}(B), \mathrm{ub}(A) + \mathrm{ub}(B)],\\
A - B &= [\mathrm{lb}(A) - \mathrm{ub}(B), \mathrm{ub}(A) - \mathrm{lb}(B)],\\
A \cdot B &= [\min\{\mathrm{lb}(A)\cdot\mathrm{lb}(B), \mathrm{lb}(A)\cdot\mathrm{ub}(B), \mathrm{ub}(A)\cdot\mathrm{lb}(B), \mathrm{ub}(A)\cdot\mathrm{ub}(B)\},\\
&\qquad \max\{\mathrm{lb}(A)\cdot\mathrm{lb}(B), \mathrm{lb}(A)\cdot\mathrm{ub}(B), \mathrm{ub}(A)\cdot\mathrm{lb}(B), \mathrm{ub}(A)\cdot\mathrm{ub}(B)\}],\\
A/B &= A \cdot [1/\mathrm{ub}(B), 1/\mathrm{lb}(B)], \quad \text{if } 0 \notin B.
\end{aligned}$$

Let $\varphi : S \subseteq \mathbb{R} \to \mathbb{R}$ be a real *standard function*, which is continuous on all $A \in \mathbb{I}(S)$. The interval extension of φ is defined by $\Phi : \mathbb{I}(S) \to \mathbb{I}, \Phi(A) := \{\varphi(a) \mid a \in A\}$. The interval extension of a given standard function is usually determined from its monotonicity properties. For example, using the fact that the exponential function e^x is strictly monotonically increasing on $\mathbb{R}$, its interval extension $E : \mathbb{I} \to \mathbb{I}$ can be defined as

$$E^A := [e^{\mathrm{lb}(A)}, e^{\mathrm{ub}(A)}], \quad A \in \mathbb{I}.$$

A vector of n intervals is called an *n-dimensional interval* (or simply a *box*): $X = (X_1, X_2, \ldots, X_n)$, $X \in \mathbb{I}^n$, and $X_i \in \mathbb{I}$ for $i = 1, 2, \ldots, n$. Moreover, for a given set of n-dimensional vectors $S \subseteq \mathbb{R}^n$, the set of n-dimensional boxes in S is denoted by $\mathbb{I}(S)$ (which is similar to the one-dimensional case). The extension of operations and functions for multidimensional intervals is defined componentwise, as it is for real vectors. The width of the n-dimensional interval vectors is defined by $w(X) = \max_i\{w(X_i)\}$, $X \in \mathbb{I}^n$.

In order to define interval extensions for compound real functions, we introduce the concept of *interval inclusion functions*. We call $F : \mathbb{I}(S) \to \mathbb{I}$ an inclusion function of $f : S \subseteq \mathbb{R}^n \to \mathbb{R}$, if $f(X) := \{f(x) \mid x \in X\} \subseteq F(X)$ holds for all $X \in \mathbb{I}(S)$. In the previous definition $f(X)$ denotes the *range of f over X*. In other words, F is an inclusion function of f, if $x \in X$ implies $f(x) \in F(X)$ for all $X \in \mathbb{I}(S)$. One of the possible ways of constructing such interval functions is the *natural interval extension*: in the real-type function expression the variables are replaced by intervals, and the operators and elementary functions are replaced by their interval analogues.

An important property of the inclusion functions is the *inclusion isotonicity* (or inclusion monotonicity) . F is said to be inclusion isotone, if $F(X) \subseteq F(Y)$, whenever $X \subseteq Y$. For example, inclusion functions constructed by the natural interval extension are isotone.

In accordance with the above definitions, from now on we shall use the following notational conventions: real numbers, vectors, and real-type functions are denoted by lower-case letters, while upper-case letters are reserved for intervals, boxes, and inclusion functions.

The statement

$$f(X) \subseteq F(X) \tag{6.4}$$

is often called the central definition of interval arithmetic. For the basic arithmetic operators and for the continuous standard functions, the enclosure $F(X)$ equals the exact range. For compound functions, however, in most cases the interval evaluation yields an interval enclosure wider than the exact range. An important research direction of interval analysis is the development of techniques that reduce this kind of *overestimation.* (A special case when an arithmetic expression can be evaluated without overestimation is the *single use expression* or SUE. In a SUE, each variable occurs exactly once. For instance, the expression of computing the Euclidean distance between two points is a SUE, but the objective function (6.3) is not.)

Besides the theoretical reliability of interval computations, the inclusion properties should also be guaranteed in computer implementations of a problem. That is, interval operations should be handled to control rounding errors. This is usually done by the interval arithmetics libraries, using exactly representable floating-point numbers (also called *machine numbers*) as the bounds of the intervals, and applying directed *outward rounding*: the bounds of the result of each interval operation are computed by properly switching between the upward and downward rounding floating-point modes. Note that due to the outward roundings the inclusion sharpness of the interval operators and the elementary functions may be lost, but the central inclusion property (6.4) still remains valid. This is a key feature of the problem solving procedure.

6.3 A prototype interval global optimization algorithm

Due to its ability to bound ranges of functions, natural interval extension provides an efficient and easily applicable tool for implementing bounding procedures for rectangular branch–and–bound global optimization algorithms. In fact, interval B&B methods were one of the first fields of application for interval arithmetics.

In this section we introduce an interval branch–and–bound method for computing all the global maximizers and the f^* maximum value for the global optimization problem

$$\max_{z \in Z_0} f(z), \tag{6.5}$$

where $f : \mathbb{R}^n \to \mathbb{R}$ is a continuous objective function for which an inclusion function exists, and $Z_0 \in \mathbb{I}^n$ is the search box.

Algorithm 5 : An interval B&B global optimization model algorithm

Inputs: – F: an inclusion function of the the objective function,
– Z_0: the search box,
– ε: tolerance value for the stopping criterion.

Outputs: – $Maximum$: enclosure of the global maximum value,
– $ResultList$: set of candidates for containing the global maximizers.

1. $Z := Z_0$; $WorkList := \{(Z, \mathrm{ub}(F(Z)))\}$; Set an initial $\tilde{f}$ cutoff value.
2. **while** ($WorkList$ is not empty)
3. $(Z, \mathrm{ub}(F(Z))) := Head(WorkList)$;
4. $Delete(Head(WorkList))$;
5. $Bisection(Z, U^1, U^2)$;
6.   **for** ($i := 1$ **to** 2)
7.   Try to improve $\tilde{f}$;
8.   Apply accelerating devices for U^i;
9.   **if** (U^i can be deleted as a whole) **then** **continue** with the next i;
10.   **endif**
11.   **if** $((w(F(U^i)) < \varepsilon))$ **then** $Insert(ResultList, (U^i, \mathrm{ub}(F(U^i))))$;
12.   **else** $Insert(WorkList, (U^i, \mathrm{ub}(F(U^i))))$;
13.   **endif**
14.   **endfor**
15. **end while**
16. $Maximum := (\tilde{f}, \max\{\mathrm{ub}(F(Z)) \mid (Z, \mathrm{ub}(F(Z))) \in ResultList\})$;
17. **return** ($Maximum$, $ResultList$);

The pseudo-code of the algorithm is given in Algorithm 5. During the execution of the algorithm an interval inclusion function $F(Z)$ of $f(z)$ is used. For each iteration cycle (between Step 2 and Step 15 of Algorithm 5), we choose a box Z from the boxes stored in the 'WorkList', and split it into two parts U^1 and U^2 (Step 5). Then for both U^i we try to delete those parts of U^i which cannot contain a global optimizer point (Steps 8 and 9). If the remaining part of U^i (also denoted by U^i in the algorithm) fulfills the *termination criterion*, we put it into the 'ResultList' (Step 11). Otherwise we store the remaining part of U^i for further processing (Step 12).

We store the boxes satisfying the stopping criterion of Step 11 in the single-linked list ResultList, where each box Z is stored together with $\mathrm{ub}(F(Z))$. The ResultList is organized according to the FIFO (first in, first out) insertion strategy.

The structure WorkList is used to store boxes which require further processing (subdivision). In our first optimization algorithm described in Chapter 7, a single-linked list is used for this purpose, the elements of which being sorted in decreasing order of the $\mathrm{ub}(F(Z))$ values. Elements with the same $\mathrm{ub}(F(Z))$ values are stored one after another, chronologically sorted, as in the ResultList.

To improve the performance of the list operations, in the advanced algorithm of Chapter 8, the WorkList is implemented as a balanced binary search tree, namely an

AVL-tree. The tree elements are sorted again in decreasing order of the $\text{ub}(F(Z))$ values. Elements having the same $\text{ub}(F(Z))$ values are stored at the same node of the tree in a single-linked list. The latter list type has the same properties as that of the ResultList.

The implementation of the WorkList provides the functions *Insert()* for insertion, *Delete()* for deletion, and *Head()* for returning the first element. Since we always select the first element of the WorkList for subdivision, the sorting criteria applied also determine the *interval selection rule*. In the present case, we choose the box with the largest $\text{ub}(F(Z))$ value; this rule is called the *Moore–Skelboe selection rule*.

In order to implement a fully functional interval B&B algorithm, there are some other things that have to be specified. These include the evaluation of $F(Z)$, the bisection strategy (Step 5), and the accelerating devices (Step 8). In the next sections we will discuss these in more detail.

6.4 An interval inclusion function for the objective function of the point arrangement problem

In the case of n points, the objective function (6.3) of Problem 6.2 is a $2n$-dimensional function. The interval inclusion of a real-type feasible solution $(x, y) \in [0, 1]^{2n}$ is given as $(X, Y) \subseteq [0, 1]^{2n}$, where $x_i \in X_i,\ y_i \in Y_i,\ i = 1, 2, \ldots, n$. Such an inclusion can be visualized as a number of n rectangles in the unit square, each rectangle containing one of the points to be packed. The theorem below gives a non-trivial inclusion function for the special form of the objective:

Theorem 6.3. *Let* $(X, Y) \subseteq [0, 1]^{2n}$*, and let*

$$D_{ij} = (X_i - X_j)^2 + (Y_i - Y_j)^2, \qquad \forall i, j \in \{1, 2, \ldots, n\}.$$

Then an inclusion function of $f_n(x, y)$ *on the* $2n$*-dimensional box* (X, Y) *is given by*

$$F_n(X, Y) := [\min_{1 \leq i \neq j \leq n} \text{lb}(D_{ij}), \min_{1 \leq i \neq j \leq n} \text{ub}(D_{ij})].$$

PROOF. We show that for the defined $F_n(X, Y)$ the required inclusion property holds: $f_n(x, y) \in F_n(X, Y)$ whenever $(x, y) \in (X, Y)$. Let $((x_1, \ldots, x_n), (y_1, \ldots, y_n)) = (x, y)$ be arbitrarily chosen from (X, Y). Using the inclusion property of the basic arithmetic operations and elementary functions (see Section 6.2), $d_{ij} \in D_{ij}$ holds for all $1 \leq i \neq j \leq n$, and therefore

$$\text{lb}(D_{ij}) \leq d_{ij} \leq \text{ub}(D_{ij}), \qquad \forall 1 \leq i \neq j \leq n.$$

Denote the index pair for which d_{ij} is minimal by $(\tilde{i}, \tilde{j})$. Should we have several index pairs satisfying this criterion, choose one of them arbitrarily. With this notation, we obtain

$$\min_{1 \leq i \neq j \leq n} \text{lb}(D_{ij}) \leq \text{lb}(D_{\tilde{i}\tilde{j}}) \leq d_{\tilde{i}\tilde{j}} = \min_{1 \leq i \neq j \leq n} d_{ij}. \qquad (6.6)$$

Now denote the index pair for which $\mathrm{ub}(D_{ij})$ is minimal by $(\hat{\imath}, \hat{\jmath})$. Again, if we have several pairs like this one, choose one of them arbitrarily. Similar to the above relations, we get

$$\min_{1\leq i\neq j\leq n} d_{ij} \leq d_{\hat{\imath}\hat{\jmath}} \leq \mathrm{ub}(D_{\hat{\imath}\hat{\jmath}}) = \min_{1\leq i\neq j\leq n} \mathrm{ub}(D_{ij}). \tag{6.7}$$

Finally, the leftmost and rightmost expressions of (6.6) and (6.7), respectively, result in

$$\min_{1\leq i\neq j\leq n} \mathrm{lb}(D_{ij}) \leq \min_{1\leq i\neq j\leq n} d_{ij} \leq \min_{1\leq i\neq j\leq n} \mathrm{ub}(D_{ij}),$$

which is exactly what is required for the interval inclusion property. □

6.5 The subdivision step

Bisection step. In Step 5 of Algorithm 5, we split the box Z perpendicular to its widest component into two boxes. If two or more components have the same width, we choose the one which has the smallest index. This is the classical subdivision method. Recently the efficiency of some more sophisticated rules (concerning the subdivision direction selection and the number of resulting subboxes) has been investigated for general interval B&B procedures—see [15, 17, 71]. For example the insertion of these rules in the respective algorithms for circle packing problems could be the subject of a future study. However, as we will see, the number of iterations is quite limited for each execution of the advanced B&B algorithm described in Chapter 8, hence the effect of changing the subdivision rule is expected to be smaller than that seen for the more general problems.

6.6 Accelerating devices

In Step 8 of Algorithm 5 several tests can be performed to delete those parts of U^i which cannot contain global maximizers. As a special case, even the whole box can be rejected on occasion. In order to test whether the investigated regions contain a global maximizer, we assume that a guaranteed lower bound $\tilde{f} \in \mathbb{R}$ for the global maximum value exists, and is available. In general, $\tilde{f} := \mathrm{lb}(F(Z_0))$ can be computed, say. For practical considerations we will use the following notation:

$\tilde{f}_0$ for the best known lower bound for the maximum of Problem 6.1 (point arrangement with distances), and

$\tilde{f}$ for the best known lower bound for the maximum of Problem 6.2 (point arrangement with squared distances), respectively.

In the present algorithms the initial $\tilde{f}$ value was determined as follows: First of all, we performed an interval function evaluation (with natural interval extension) on a tiny interval around the machine number representation of the currently best-known packing, and set $\tilde{f}$ to the lower bound of the result interval. The corresponding $\tilde{f}_0$ was obtained by an interval square root function evaluation:

$$F_s := \sqrt{\tilde{f}} \in \mathbb{I}, \quad \tilde{f}_0 := \mathrm{lb}(F_s). \tag{6.8}$$

Below we discuss the accelerating tests applied to *both* the first and the advanced version of the optimization method, described in Chapter 7 and Chapter 8, respectively. The different accelerating tools (or their different implementations) will be discussed when we describe the particular methods.

6.6.1 The cutoff test

After computing the inclusions $F(U^i)$ of the objective function over the boxes U^i $(i = 1, 2)$, U^i can be eliminated if $\mathrm{ub}(F(U^i)) < \tilde{f}$. Note that if $\tilde{f}$ is improved in Step 7, then one can discard all those $(Z, \mathrm{ub}(F(Z)))$ elements in the WorkList, for which $\mathrm{ub}(F(Z)) < \tilde{f}$ holds. As experience shows, updating $\tilde{f}$ can be a very hard task for a point arrangement problem, requiring the verified results of some sophisticated search methods (e.g. those in [11, 36, 86]). Nevertheless, when the goal is to verify the optimality of a given packing, the use of an appropriate initial $\tilde{f}$ might be enough.

6.6.2 The 'method of active regions'

This accelerating test is known from the literature (see e.g. [87, 88, 94]) as a part of non-interval type methods. The idea behind this approach is the following: assume that we have a reliable $\tilde{f}_0$ value. Consider the closed, two-dimensional regions R_i, $i = 1, \dots, n$ within the unit square, where R_i should contain the ith point to be placed. Then from each R_i we can delete those points that have a distance smaller than $\tilde{f}_0$ from *all* points of another region R_j, $i \neq j$.

A basic algorithm of the method of active regions is given in Algorithm 6. Consider a box $(X, Y) \subseteq [0, 1]^{2n}$. The rectangle $(X_i, Y_i) \subseteq [0, 1]^2$ is called the ith *initial active region*, $i = 1, \dots, n$. During the procedure, the R_i active regions of the various components are reduced iteratively, until either one of the active regions becomes empty or a preset iteration limit ($It_{\max}$) is reached. In the first case the whole box (X, Y) can be deleted (Step 5). In the latter case a new box (X', Y') containing the remaining regions will be stored (Steps 9 and 10). The most important part of Algorithm 6 is Step 4, in which we delete those points (forming an *inactive region*) of R_i that have a distance smaller than $\tilde{f}_0$ from all points of R_j.

One crucial part of the algorithm is the representation of the intermediate active areas (i.e. the R_i regions). One can easily show that a set of points within a two-dimensional geometrical object having a distance at least $\tilde{f}_0$ to all points of another object may be non-convex or even non-connected. In any case, a good approximation of the active and inactive point sets is vital to remove as large inactive sets as possible.

Algorithm 6 : A prototype algorithm for the method of active regions

Inputs: – $\tilde{f}_0$: a validated lower bound for the global maximum of Problem 6.1,
– $(X, Y) \subseteq [0, 1]^{2n}$: the box to be reduced,
– $It_{\max}$: the iteration limit.

Output: – $(X', Y') \subseteq [0, 1]^{2n}$: a box containing the remaining regions.

for ($i := 1$ **to** n) **do** $R_i := (X_i, Y_i)$;
for ($i := 1$ **to** $It_{\max}$) **do**
  forall ((i, j), $1 \leq i, j \leq n$, $i \neq j$) **do**
    $R_i' := Diminish\_ij((R_i, R_j), \tilde{f}_0)$;   *comment: reduce R_i to R_i'*
    if ($(R_i' = \emptyset)$) **then** **return** "(X', Y') is empty";
    $R_i := R_i'$;
  endfor
endfor
for ($i := 1$ **to** n) **do** $(X_i', Y_i') := Rectangular\_enclosure(R_i)$;
return (X', Y');

In [94], the initial active regions were quantized into many rectangular pieces, applying splittings both in horizontal and in vertical directions—and the set of eliminated and remaining pieces representing the inactive and active point sets, respectively. In [88], a similar approach was applied, but with splittings only in one direction. Until now, the most efficient realization is the one by Nurmela and Östergård [87], which approximated the active and inactive regions by polygons.

Obviously, for an interval method the rectangular approach is more appropriate to use. That is why in our first method we implemented an interval version similar to the one in [88]. It allows us to choose between the horizontal and rectangular splittings, and perform only one of them. This algorithm variant will be discussed in Section 7.3. In later phases of our studies, however, we suspected that the additional effort of implementing an interval-based polygon approximation algorithm would improve the performance of our advanced method substantially. As it turned out, one of the key algorithmic improvements for the solution of the packing problem of 28, 29, and 30 circles was the interval method of active regions using polygons. This algorithm will be elaborated in Section 8.1.

7

The First Fully Interval-based Optimization Method

As we mentioned in Chapter 6, during our investigations we designed two versions of a fully interval-based optimization method to solve point packing problems in a reliable and systematic way. In this chapter we introduce our initial method, which serves as a starting point for the advanced algorithm. The usual techniques applied in the interval B&B algorithm frame were already discussed in the previous chapter. Below we specify three special accelerating methods, designed specifically for the point arrangement problem class. The first one is based on the monotonicity properties of the objective function, the second is a simple test to detect the free points (corresponding to free circles) of a packing configuration, and finally, the third is a rectangular implementation of the method of active areas. Then we present the first numerical results related to the local and global verification of the currently existing optimal solutions using interval calculations.

7.1 A monotonicity test for the point arrangement problems

One of the most efficient accelerating devices of the standard interval B&B methods is the monotonicity test , which is used to exclude suboptimal parts of the search region by utilizing the monotonicity of the objective function. For continuously differentiable functions, the monotonicity test is carried out by evaluating the interval enclosure of the gradient (see [43], say). However, it is easy to see that due to the min() function, the objective function (6.3) is not differentiable (and thus, not continuously differentiable) on the whole search domain $[0,1]^{2n}$. On the other hand, since (6.3) is obviously monotonic in certain regions, it is worth designing a special tool that explores and utilizes this property. To achieve this goal, we need to look more closely to investigate the geometry of the minimal pairwise distance computations.

The following theorem provides a monotonicity test for (6.3) without gradient evaluations:

Theorem 7.1. *Let* $(X,Y) \in \mathbb{I}^{2n}$ *be the actual box while Algorithm 5 is running. When for some* k ($k \in \{1,2,\ldots,n\}$), *and for* $j \in \{1,2,\ldots,n\}, j \neq k$ *one of the*

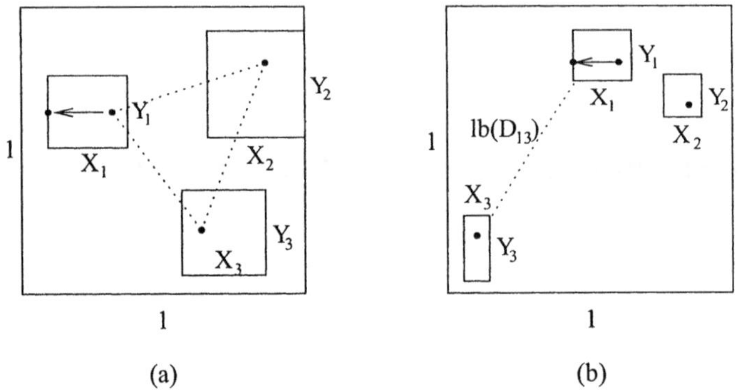

Fig. 7.1. Monotone decreasing property in variable x_1. In the first case, the (7.1) assumption holds for $j = 2, 3$. In the second case, (7.1) holds for $j = 2$, while (7.2) holds for $j = 3$.

conditions:

$$\mathrm{ub}(X_k) \leq \mathrm{lb}(X_j), \text{ or} \tag{7.1}$$

$$\mathrm{lb}(D_{kj}) > \mathrm{ub}(F_n(X, Y)) \tag{7.2}$$

holds, then f_n is monotonically decreasing on the box (X, Y) in the variable x_k, and thus, X_k can be shrunk to the point interval $[\mathrm{lb}(X_k), \mathrm{lb}(X_k)]$.

PROOF. Let $(x, y) \in \mathbb{R}^{2n}$ be arbitrarily chosen from (X, Y). We show that the real function value $f_n(x, y)$ will not decrease while x_k is decreasing within X_k. Figure 7.1 shows two typical cases for $n = 3$ (i.e. for three rectangles in the unit square) and for $k = 1$. Based on what we said in the previous chapter, the actual box is represented by the three rectangles (X_i, Y_i), and a feasible solution can be obtained by selecting one point from each rectangle.

(i) When (7.1) holds for some j (where $j \in \{1, 2, \ldots, n\}, j \neq k$), then the squared distance between (x_j, y_j) and (x_k, y_k) will not decrease while x_k is decreasing. This situation is shown in Figure 7.1a, Figure 7.1b for $j = 2$, and in Figure 7.1a for $j = 3$.

(ii) For the other j indices which do not satisfy (7.1), the distance between (x_j, y_j) and (x_k, y_k) can decrease after x_k has decreased (see (X_3, Y_3) in Figure 7.1.b). However, for these j indices, the squared distance between the corresponding points must be still greater than $\mathrm{lb}(D_{kj})$, which is, from (7.2), greater than the upper bound of the interval function value. In other words, the squared distance of the point pair selected from the jth and kth rectangle is always out of the range of $F_n(X, Y)$—that is, the change of d_{jk} has no effect on the objective function value.

(iii) Finally, the squared distance between any two point pairs $(x_p, y_p), (x_q, y_q)$, $p, q \neq k$ will remain constant while x_k is varying.

Summarizing items (i), (ii), and (iii), we see that the minimal squared distance between the points will be greater than or equal to when x_k is decreased in X_k. Equivalently, f is monotonically decreasing on (X, Y) in x_k. □

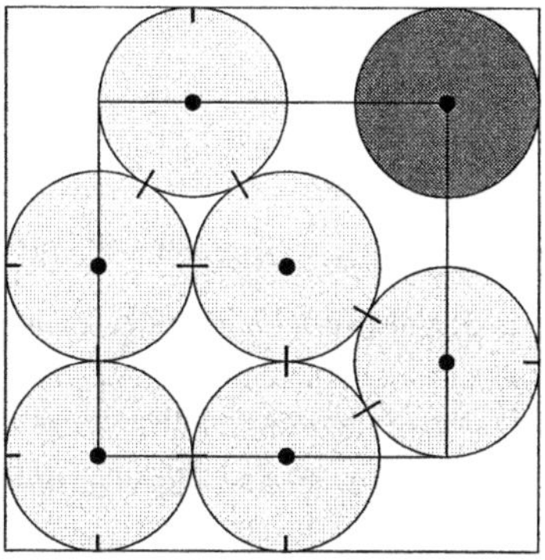

Fig. 7.2. The optimal packing of seven circles.

Remark 7.2. Evidently, the role of X and Y in Theorem 7.1 can be swapped. Furthermore, a similar theorem can be formulated for the monotonically increasing case. During a practical application, all $2n$ variables of Theorem 7.1 are made use of, testing both the decreasing and the increasing case.

Remark 7.3. It is worth bearing in mind that we have proven only monotonicity and not strict monotonicity on the actual box. This means that we do not necessarily obtain *all* optimizers when using the above accelerating test.

One big advantage of the monotonicity test is that some points in the optimal solution will be placed on the side of the unit square. By applying Theorem 7.1, we can explore these points in the early stages of the B&B algorithm, and reduce the search area in a very efficient way.

7.2 A special case of monotonicity—handling the free points

One of the efficiency improvement techniques comes from examining the 'free-circles' phenomenon where a number of equivalent optimizer points obviously exists. (See [86, 87, 94].) To illustrate this property, let us consider the example illustrated in Figure 7.2, which shows the optimal packing of seven circles. Here the location of one of the circles is not fixed; it can be moved while retaining the optimal radius.

Note that by the equivalence of the point arrangement and circle packing problems (Theorem 2.9), each free point in an optimal point arrangement is identified by the centre of a *free circle* of the corresponding optimal packing of circles, and vice versa. Formally, we define a free point of the *optimal points arrangement* configuration in the following way:

Definition 7.4. *Consider the point arrangement problem class with the distance function and objective function given by (6.3). Consider an optimal point arrangement, i.e. a vector* $(x, y) = (x_1, \ldots, x_n, y_1, \ldots, y_n) \in \mathbb{R}^{2n}$*, for which* f_n *is maximal. We call a point* $p_k = (x_k, y_k)$, $k \in \{1, \ldots, n\}$ *of this optimal packing a* free point, *or* *equivalently,* a free circle centre, *if there exists a half line H with the endpoint p_k and there exists a positive number ε, such that*

$$f_n(x,y) = f_n(x_1,\ldots,x'_k,\ldots,x_n,y_1,\ldots,y'_k,\ldots,y_n)$$

for all $(x'_k, y'_k) \in H \cap N_\varepsilon(p_k)$, where $N_\varepsilon(p_k)$ denotes the ε-neighbourhood of p_k.

The meaning of the above definition is straightforward: we say that a point is free in an optimal packing if this point can be slightly shifted without losing the optimality of the packing. It is also obvious that the handling of free points is crucial when solving point or circle packing problems, since they form a continuum (and usually a positive measure set) of equivalent global optimizers.

It is easy to see that the occurrence of free points is a special case of monotonicity, since the objective function remains constant while moving a free point p_k along $H \cap N_\varepsilon(p_k)$. Based on Theorem 7.1, the next theorem examines the case when f_n is constant *while moving p_k arbitrarily within its enclosing rectangle* $(X_k, Y_k) \subseteq [0,1]^2$ (in other words, each such $p_k \in (X_k, Y_k)$ is a free point):

Theorem 7.5. *Let $(X,Y) \in \mathbb{I}^{2n}$ be the actual box while Algorithm 5 is running. When for some k ($k \in \{1,2,\ldots,n\}$), and for all $j \in \{1,2,\ldots,n\}, j \neq k$,*

$$\mathrm{lb}(D_{kj}) > \mathrm{ub}(F_n(X,Y)) \tag{7.3}$$

holds, then for all optimal point arrangements $(x,y) \in (X,Y)$ the point (x_k, y_k) is a free point, and it can be moved arbitrarily in (X_k, Y_k) without losing optimality. Thus, for later investigations of the box (X,Y) it is sufficient to consider (x_k, y_k) for (X_k, Y_k) together while bearing in mind that 'each point of (X_k, Y_k) is a free point'.

PROOF. Let $(x,y) \in (X,Y)$ be an optimal point arrangement. Then

$$f_n(x,y) \leq \mathrm{ub}(F_n(X,Y)) < \mathrm{lb}(D_{kj}).$$

Consequently, the location of (x_k, y_k) within (X_k, Y_k) does not affect the objective function value, that is,

$$f_n(x,y) = f_n(x_1,\ldots,x'_k,\ldots,x_n,y_1,\ldots,y'_k,\ldots,y_n)$$

for each $(x'_k, y'_k) \in (X_k, Y_k)$. □

The biggest drawback of the above test is that the set of free points of the optimal arrangements is usually not representable by rectangles (or by a finite set of rectangles). Instead, they are determined by a set of arcs as well because of the neighbouring points of the packing. This is another reason why the previously mentioned 'method of active regions' is necessary, the implementation of which will be introduced in the following section.

An improved version of the method of handling free points will be a part of our advanced optimization method, and will be discussed in Section 8.3.

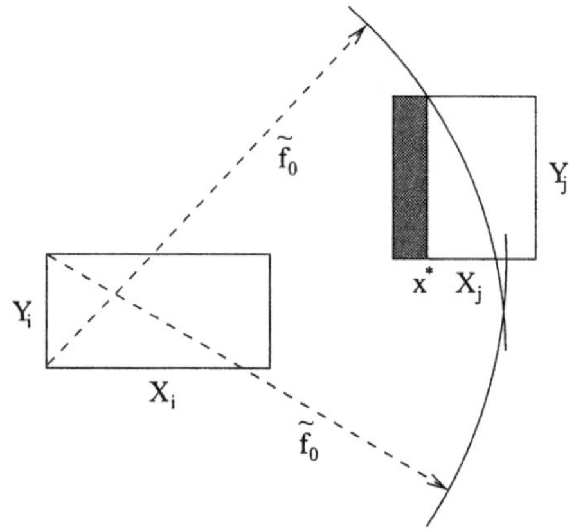

Fig. 7.3. A special case of the method of active regions. The shaded rectangle can be deleted.

7.3 A rectangular approach for the 'method of active regions'

The basic idea behind the method of active regions was given in Section 6.6.2. In the following, we introduce an interval-based implementation of the elementary region elimination method (Step 4 of Algorithm 6). We represent the active regions R_i by rectangles, i.e. we use a rectangular approximation of the exact active regions.

Figure 7.3 shows a simple case when $\mathrm{ub}(X_i) < \mathrm{lb}(X_j)$ and $\mathrm{ub}(Y_i) < \mathrm{lb}(Y_j)$. Here the inactive region includes those points which have a distance less than $\tilde{f}$ to $(\mathrm{lb}(X_i), \mathrm{lb}(Y_i))$. This way the shaded part of (X_j, Y_j) can be deleted as an inactive region. In our interval implementation we need to have rectangles as the result of the elimination step.

When considering rectangular approximations, the core procedure $Diminish\_ij()$ of the method of active areas is based on solving the following problem:

Consider the (X_i, Y_i), $(X_j, Y_j) \subseteq [0,1]^2$ *two-dimensional boxes. Assume that*

(i) the interior of $(X_i, Y_i) \cap (X_j, Y_j)$ *is empty and, furthermore,*
(ii) $\mathrm{ub}(X_i) \leq \mathrm{lb}(X_j)$.

We look for the largest *possible* $x^* \in \mathbb{R}$, *for which*

$$\mathrm{ub}((X_i - [\mathrm{lb}(X_j), x^*])^2 + (Y_i - Y_j)^2) \leq \tilde{f}_0^2, \tag{7.4}$$

that is, the distance between any points of $([\mathrm{lb}(X_j), x^*], Y_j)$ *and* (X_i, Y_i) *is at most* $\tilde{f}_0$. *In other words, the rectangle* $([\mathrm{lb}(X_j), x^*], Y_j)$ *is an inactive region, hence it can be eliminated.* (In Figure 7.3, this inactive region is represented by the dark shaded area.)

Note that in (7.4) the equality case is permitted, as the upper bound of the eliminated region is the same as the lower bound of the remaining region; in other words, both regions are closed. That is why it is important to mention both the inactive and the remaining rectangle: if in a special case $x^* = \text{ub}(X_j)$, then the whole original rectangle (X_j, Y_j) can be eliminated in the above manner, but still, the line segment $([\text{ub}(X_j), \text{ub}(X_j)], Y_j)$ has to be considered as a remaining region. In the polygon-based variant of this method (Section 8.1), a slightly different approach is used, which helps to avoid the above special case. In that algorithm the remaining regions are determined by an $\tilde{f}_l$ value with $\tilde{f}_l < \tilde{f}_0$.

It is also worth remarking that the two prerequisites (i) and (ii) will not introduce any new restriction: (i) will be guaranteed by the *tiling* method introduced later in this chapter. If (i) holds, then condition (ii) can be also guaranteed by temporarily swapping the x and y components, if necessary. (The latter condition is needed only to fix the relative position of the two rectangles.)

Figure 7.3 also shows that in a general situation the inactive region is determined by the two arcs drawn from the points $(\text{lb}(X_i), \text{lb}(Y_i))$ and $(\text{lb}(X_i), \text{ub}(Y_i))$, and by the lines $y = \text{ub}(Y_j)$ and $y = \text{lb}(Y_j)$, respectively. The mathematical explanation for this is based on a simple, but important statement of Nurmela and Östergård:

Lemma 7.6. *Nurmela and Östergård [87]: If a point has a distance smaller than $\tilde{f}_0$ from each vertex of a polygon, then it has a distance smaller than $\tilde{f}_0$ from every point of the polygon.*

Returning to our special case, consider the equations

$$(\text{lb}(X_i) - x^*)^2 + (\text{lb}(Y_i) - \text{ub}(Y_j))^2 = \tilde{f}_0^2, \tag{7.5}$$

and

$$(\text{lb}(X_i) - x^*)^2 + (\text{ub}(Y_i) - \text{lb}(Y_j))^2 = \tilde{f}_0^2. \tag{7.6}$$

Let $x_1^* \in \mathbb{R}$ and $x_2^* \in \mathbb{R}$ be the solution of (7.5) and (7.6), respectively (when they exist).

Assuming *exact computations*, x_1^* can be determined as follows. In (7.5), if $\tilde{f}_0^2 - (\text{lb}(Y_i) - \text{ub}(Y_j))^2 < 0$, then x_1^* does not exist. Otherwise, using $x_1^* \geq \text{lb}(X_i)$, we obtain

$$x_1^* = \sqrt{\tilde{f}_0^2 - (\text{lb}(Y_i) - \text{ub}(Y_j))^2} + \text{lb}(X_i).$$

Similarly, if x_2^* exists, then it is given by

$$x_2^* = \sqrt{\tilde{f}_0^2 - (\text{ub}(Y_i) - \text{lb}(Y_j))^2} + \text{lb}(X_i).$$

Consequently, if neither x_1^* nor x_2^* exists, then the *whole* (X_j, Y_j) rectangle can be eliminated, and the remaining region is empty. If exactly one of the x_k^* $(k = 1, 2)$ exists, then $x^* := x_k^*$, otherwise x^* can be set to $\min(x_1^*, x_2^*)$.

Since we have to implement the above method on a computer, we have to carry out all the computations in a reliable way, using interval calculations. That is, we need to compute the interval enclosures of $X_1^*, X_2^* \in \mathbb{I}$ of the exact x_1^* and x_2^* values, and then determine a guaranteed lower bound for x^* as a machine number. This will guarantee that we eliminate exclusively inactive points during the execution of the method.

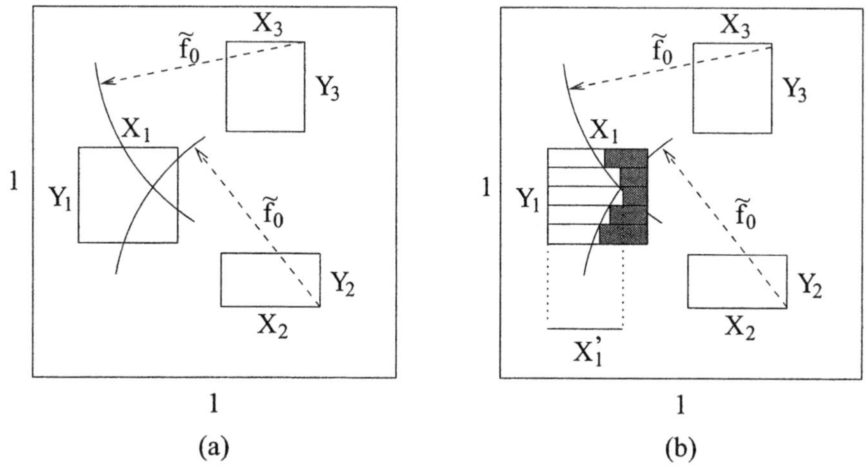

Fig. 7.4. An illustration of the method of active areas with temporary one-directional splittings.

Notice that because of the B&B frame algorithm, the bounds of the intervals X_i, X_j, Y_i, and Y_j, and the $\tilde{f}_0$ value are all machine numbers, i.e. their interval enclosures will be point intervals. From (7.5), if the upper bound of the enclosure $Z_1 = \tilde{f}_0^2 - (\mathrm{lb}(Y_i) - \mathrm{ub}(Y_j))^2 \in \mathbb{I}$ is smaller than 0, then X_1^* does not exist. Otherwise, we can have $Z_1 := (\max\{0, \mathrm{lb}(Z_1)\}, \mathrm{ub}(Z_1))$, and after the interval square root operation we get

$$X_1^* = \sqrt{Z_1} + \mathrm{lb}(X_i).$$

Similarly, $Z_2 := \tilde{f}_0^2 - (\mathrm{ub}(Y_i) - \mathrm{lb}(Y_j))^2 \in \mathbb{I}$, and if X_2^* exists, then it is given by

$$X_2^* = \sqrt{Z_2} + \mathrm{lb}(X_i).$$

Finally, if neither X_1^* nor X_2^* exists, then the *whole* (X_j, Y_j) can be deleted, and the remaining region is empty. If exactly one of the X_k^* intervals exists, then $x^* := \mathrm{lb}(X_k^*)$; otherwise $x^* := \min(\mathrm{lb}(X_1^*), \mathrm{lb}(X_2^*))$.

After obtaining x^*, the inactive and remaining regions can be computed in a straightforward way: if $x^* > \mathrm{ub}(X_j)$, then the whole (X_j, Y_j) can be eliminated, and the remaining region is empty. If $x^* < \mathrm{lb}(X_j)$, then we cannot eliminate any points, that is, the remaining region is (X_j, Y_j). In all the other cases we can eliminate the rectangle $([\mathrm{lb}(X_j), x^*], Y_j)$. Due to the interval computations that are performed, the method ensures that the remaining active region contains the region that would be obtained by exact computations. Thus we do not lose any active points. This is a very important feature of the method.

The above procedure itself is viable when applied to the component rectangles of the U^i box of Algorithm 5. In order to achieve further performance improvements, we included another additional useful idea into the rectangular 'method of active regions': when determining active areas, we *temporarily split* all rectangles into some pieces in one direction—that is, each R_i region of Algorithm 6 will be a *union* of smaller rectangular pieces. Depending on the relative location of the rectangles, our algorithm can choose between horizontal or vertical splitting.

The idea of subdividing the current search area in similar ways is not new; the technique of splitting each rectangle into many pieces in both directions was introduced in [94], which was then further developed by the same authors to apply splittings in only one direction.

To illustrate the advantage of this temporary splitting, let us consider Figure 7.4. In this particular case, we try to delete the inactive points from (X_1, Y_1). As you will notice, by not splitting this rectangle into some subregions, neither (X_2, Y_2) nor (X_3, Y_3) can reduce X_1 or Y_1 (Figure 7.4a). But after splitting Y_1 into five subintervals, the X_1 components of the five rectangles can be reduced (see the shaded areas in Figure 7.4b), using the two reducing rectangles *at the same time*. Consequently, a large part of the whole (X_1, Y_1) can be eliminated (X_1 can be reduced to X_1').

Remark 7.7. An important parameter of the temporary splitting technique is the number of generated smaller rectangles. Obviously, the bigger this number is, the better the approximation of the active regions will be, but the computational complexity also increases. As our experiences show, splitting into 20 subregions fulfils our expectations on the first interval method for point arrangement. On the other hand, it is also possible to implement a kind of adaptive splitting. Such an adaptive technique would dynamically change the number of subregions, i.e. the quality of the approximation during the execution of the B&B algorithm.

An advanced version of the method of active regions, namely the interval implementation of the method of [87] using polygon approximations, will be elaborated on in Section 8.1.

7.4 Numerical results

The results of the previous sections provided us with sufficient tools for implementing a fully interval-based reliable algorithm capable of solving point arrangement problems. In this section we present some numerical results obtained with this algorithm.

The numerical tests were carried out on a PC with a 333 MHz processor and 128 MB RAM running on the Linux platform. We also made use of the C/C++ interval arithmetic library PROFIL/BIAS [54]. The basic framework of the branch-and-bound algorithm and the list handling task was done using one from the C–XSC Numerical Toolbox [43] with some modifications. This modified method was introduced in [71].

7.4.1 Local verification

Prior to our studies, real-type optimal circle (or point) packings in a unit square were known from 1 up to 27 (cf. Table 1.2 of Section 1.2.8). Some of them were proved in a theoretical way (the first six cases and the cases 8, 9, 14, and 36), but the other problems were solved using computer-aided techniques – mostly via numerical algorithms. The first goal during the development of our interval-based algorithm was the verification of the correctness and accuracy of these earlier numerical results. (We also ran our verified algorithm on cases solved in a theoretical way in order to test the efficiency of the applied techniques.)

During the verification procedures, we set $\tilde{f}_0$ to the published optimum values (usually given to 15 decimal places). From $\tilde{f}_0$, we determined the (expected) lower bound $\tilde{f}$ of the maximum of Problem 6.2 by

$$F = \tilde{f}_0^2 \in \mathbb{I}, \; \tilde{f} := \text{lb}(F).$$

The components of the optimal solutions were extended to intervals with a diameter of 0.01, using the corresponding real value as the midpoint. (These intervals were then forced to intersect $[0, 1]$ to keep the search region within the unit square.) In this sense, our aim was a *local* verification of the optimal solutions, i.e. we did not check whether these solutions are the global optima.

The specification of the elements of Algorithm 5 was the following: the inclusion function was the one constructed in Section 6.4, the subdivision of the boxes being done according to Section 6.5. The accelerating tests were the general cut-off test (Section 6.6), the monotonicity test of Section 7.1, the method of examining free points (Section 7.2), and the method of active regions with rectangular approximations (Section 7.3). The ε value used in the termination criterion of the B&B algorithm was set to 10^{-12}, that is, we looked for very tight enclosures close to the limitations of double-precision floating-point arithmetic. The CPU time was limited to two hours.

The local verification method can return two different types of answers:

- *Acceptance*: the interval enclosure of the maximum value contains the published real-type optimum and, furthermore, one of the elements of the ResultList contains the published real-type optimizer. In this case we conclude that, *within the search region*, the corresponding problem has no solution, that is better than the upper bound of the enclosure F^* of the maximum value; and, moreover, the published optimum can differ from the exact optimum by at most $w(F^*)$.
- *Rejection*: Algorithm 5 terminates with an empty WorkList and with an empty ResultList. Then the packing problem in question has no solution that can achieve the published optimum value *within the search region*.

The computational results are summarized in Table 7.1. In the columns we displayed the number of points to be packed, the dimensionality of the problem, the CPU time (in seconds), the maximal length of the WorkList (indicating the storage complexity), the number of objective function evaluations, the number of executions of the method of active regions, and the number of B&B iterations, respectively.

As the table shows, the algorithm finished its execution in almost every case, with the exception of $n = 21, 22, 26$, and 27, which could not be solved within the given time limit. As an additional test, the problem set contained a case denoted by '21 (-)', which was used to 'verify' an incorrect solution published in [66]. (For more details on this erroneous solution see the historical summary in Section 1.2.8.) The interval

Table 7.1. Local verification of the existing solutions. The columns contain the number of points (n), the dimensionality of the problem ($dim = 2n$), the CPU time (CPU, in seconds), the maximal length of the WorkList (MLL), the number of objective function evaluations (NFE), the number of executions of the method of active regions (NMA), and the number of iterations (IT).

n	dim	CPU	MLL	NFE	NMA	IT
2	4	0.01	1	4	1	1
3	6	0.11	1	36	12	8
4	8	0.05	1	24	8	8
5	10	0.09	4	126	27	24
6	12	0.69	4	142	56	28
7	14	0.74	6	116	46	23
8	16	0.41	4	114	44	23
9	18	1.19	16	116	78	39
10	20	3.98	8	317	126	63
11	22	2.52	6	270	106	54
12	24	19.96	15	1 023	407	204
13	26	778.10	726	44 969	17 641	8 908
14	28	2.71	4	143	56	29
15	30	2.37	6	146	58	29
16	32	0.88	2	42	28	14
17	34	33.04	24	1 659	662	331
18	36	50.35	20	1 627	648	324
19	38	3 637.25	1 095	82 862	32 945	16 538
20	40	4.65	4	165	66	33
21 (-)	42	1 269.68	1 902	28 755	11 496	5 750
23	46	337.09	92	7 039	2 796	1 401
24	48	5.37	4	91	60	30
25	50	3.16	2	67	40	20
36	72	16.87	8	141	96	43

algorithm rejected this solution (correctly), and in all other problem instances here the published results were accepted within the prescribed ε tolerance value.

The most important observation related to the performance indicators is that increasing the dimensionality does not necessarily mean that the local verification becomes harder. Instead, the hard cases were the ones where the published optimal packing does not follow a regular pattern, and/or contains free point(s). The running time was determined mainly by the number of calls to the method of active regions (as it is the most expensive, but it is the most efficient accelerating test), while the storage complexity proved to be surprisingly low—at least in the completed cases.

7.4.2 Global solutions using tiles

Although the above described verification method operates on a relatively wide initial search area, the most interesting problem (and our final goal) is still to find and verify solutions globally, i.e. on the whole unit square.

It is easy to see that in the global phase of the optimization a different problem occurs, namely the difficulty of handling geometrically equivalent solutions in an efficient way while performing the numerical computations. For instance, configurations that can be obtained by permutating the set points are equivalent in a geometrical sense, but they are considered as different solutions in the numerical sense. The other problem is that each solution can be transformed to symmetric, but geometrically equivalent solutions (by reflections and rotations), which also increases the number of needlessly evaluated results. In addition, most of the elimination methods, like the method of active regions, work efficiently only in the local phase, but they are less useful when the rectangles to be reduced are very large, overlap each other, and so on.

To highlight the importance of these difficulties, it is worth mentioning that if the permutation and symmetry problems had not been resolved, the current interval method would have been able to solve the original point arrangement problems only up to $n = 5$ and have required several hours of CPU time.

To cope with these problems, in [94], a global elimination technique was proposed, based on the concept of *tiling*. First assume that an $\tilde{f}_0$ lower bound for the optimum of Problem 6.1 is known. Divide the unit square into a number of s closed, non-overlapping regions (usually into rectangles), having a maximal diameter less than $\tilde{f}_0$. Then for a solution having a function value greater than or equal to $\tilde{f}$, any such region (called 'tile') can contain at most one point of the solution. This fact enables us to choose n tiles from the total number of s tiles in all possible ways, and run the optimization method for every such tile combination.

It should be mentioned that besides finding the optimal solutions, the tiling technique is also applicable in interval-based bounding investigations—that is, in cases when our goal is to determine whether the optimum of the problem is smaller than a given, approximate optimum $\hat{f}$. (We can apply such a technique e.g. when the problem is too hard to be completely solved.) In detail, if all points of all possible tile combinations can be eliminated by the B&B method using the $\hat{f}$ value (and the accompanying $\hat{f}_0$) in the accelerating tests, then we can conclude that the optimum of the problem is guaranteed to be smaller than $\hat{f}$.

The most often used method of dividing the unit square into tiles is via its subdivision into $k \times l$ uniform rectangles with horizontal and vertical splittings. Although it has not yet been proved that this technique always results in the smallest number of tile combinations, it is easy to implement, and it can be readily applied in our interval framework. In the case of a $k \times l$ subdivision, the number of tile combinations is

$$\min \left\{ \binom{k \cdot l}{n} \mid k, l \geq 1 \text{ integers}, \ (1/k^2 + 1/l^2)^{1/2} < \tilde{f}_0 \right\}.$$

Some of the tile combinations can be eliminated using symmetry properties (e.g. by the Burnside lemma), but the primary difficulty of determining optimal packings for more than 27 circles with this technique comes from the problem of exponential growth of tile permutations. For $n = 27$, the selection $k = l = 6$ results in $\binom{36}{27} \approx 9.4 \cdot 10^7$ tile combinations. Although [87] does not give the precise computation time of the solution procedure, we estimate that it took about one month of CPU time—

Table 7.2. Global verification using the tiling approach. The table includes the number of points (n), the dimensionality of the problem ($dim = 2n$), the tiling setting (Comb.), the number of tile combinations (Tiles_nr), the total CPU time (CPU, in seconds), the maximal length of the WorkLists (MLL), the total number of objective function evaluations (NFE), the total number of executions of the method of active regions (NMA), and the total number of iterations (IT).

n	dim	Comb.	Tiles_nr	CPU	MLL	NFE	NMA	IT
2	4	2x2	6	0.02	1	8	8	2
3	6	2x2	4	0.61	1	170	66	36
4	8	2x2	1	0.09	1	27	10	9
5	10	3x2	6	0.14	1	68	26	16
6	12	3x3	84	11.65	2	1 897	813	386
7	14	3x3	36	15.99	8	1 588	660	312
8	16	3x3	9	0.84	1	75	41	16
9	18	3x3	1	0.51	1	30	18	10
10	20	4x3	66	127.67	11	4 923	2 231	1 099
11	22	5x3	1 365	625.72	48	17 395	9 414	4 065
12	24	4x4	1 820	153.34	14	3 665	3 281	736
14	28	5x4	38 760	1 088.92	14	4 246	40 698	986
15	30	5x4	15 504	454.14	6	668	15 792	147
16	32	5x4	4 845	196.85	1	33	4 866	11
19	38	5x5	177 100	13 415.15	326	48 762	197 757	10 562
20	40	5x5	53 130	3 914.12	2	316	53 311	97

using only the double precision computations instead of the 4–35 times slower interval computations. In order to solve the problem of packing 28 points, we need at least a 7×6 tiling, that is, one should process some $\binom{42}{28} \approx 5.3 \cdot 10^{10}$ tile combinations.

The hardware-software environment of the global verification tests was the one outlined before Section 7.4.1. Now the time limit was set to four hours. The input values for the B&B algorithm were the same as before, with one small modification: the method of active regions was also executed on every starting search box Z_0, i.e. before performing a subdivision on it.

The results of the global verification are shown in Table 7.2. Since the verification consists of the subsequent executions of Algorithm 5, the table lists the aggregated performance indicators for each problem instance. Summarizing the results, we find that with three exceptions the cases $n = 2, \dots, 20$ were all solved within the time limit. The final result was always the *acceptance* of the published numerical results. The acceptance was interpreted in the same way as in Section 7.4.1, but this time it was valid for the whole search space ($[0, 1]^{2n}$) of the packing problem. In the unsolved cases, the elimination of one or several individual tile combinations proved the main difficulty; for $n \geq 21$, this was exacerbated by the large number of tile combinations. Nevertheless, there were some cases (e.g. $n = 16, 20$) that required a very small number of function evaluations and iterations. This shows the strength of the method of active regions even for larger initial rectangles. Also, the reader may have noticed that in several rows of the table the number of iterations is smaller than the number of tile combinations. This is the result of executing the method of active areas before applying any subdivision (i.e. before entering the main iteration loop of the B&B algorithm).

In this chapter, the first fully interval-based optimization method was described for the point arrangement problems. While developing the algorithm, our main goal was to start from the general interval branch–and–bound optimization framework, and then construct problem specific tools which utilize the geometric settings of the problem class.

Following the theoretical investigations, some preliminary numerical tests were carried out. We obtained promising results in both the local and global validation of real-type solutions. Obviously, the results presented here could be further extended—for instance by increasing the time limit, exploiting the symmetry properties of the tile combinations for the global verification part, and so on. However, in the next steps of our research studies we chose to pay more attention to constructing a more advanced algorithm based on our previous experiences. In the next chapter we will present this improved algorithm and discuss some of its principal features.

8

The Improved Version of the Interval Optimization Method

This chapter presents the advanced version of our fully interval-based branch–and–bound optimization algorithm for solving point arrangement problems. Our starting point is again the B&B algorithm framework and the generally applicable interval tools described in Chapter 6. As one of the most efficient parts of the new algorithm, an interval-based form of a previously known version of the 'method of active regions' will be introduced. This method represents the remaining areas still of interest as polygons, having calculated them in a reliable way.

As we concluded in Section 7.4.2, the most promising strategy currently for finding optimal point or circle packing configurations is the partitioning of the original problem into subproblems by dividing the unit square into *tiles*. Yet, as a result of the greatly increasing number of resulting subproblems (tile combinations), previous computer-aided methods were unable to solve problem instances above 27 circles. This chapter presents a carefully developed technique that resolves this difficulty by eliminating large groups of subproblems together. As a demonstration of the capabilities of the new algorithm, the problems of packing 28, 29, and 30 circles were solved within very tight tolerance values. Moreover, our verified procedure decreased the uncertainty in the location of the optimal packings by more than 700 orders of magnitude in every case.

The chapter is organized as follows. Section 8.1 introduces the interval polygon approximation method for eliminating inactive regions. In Section 8.2, the questions of finding global solutions are discussed and a new approach for eliminating tile combinations is investigated. In Section 8.3, we discuss a new method for handling the occurrences of free points in optimal configurations. Then in Section 8.4 our proposed multistage method is introduced and used to solve the packing problems of 28, 29, and 30 points. The numerical results presented demonstrate how well the algorithm works in the successive elimination stages.

8.1 Method of active regions using polygons

The idea behind the method of active regions has already been discussed in Section 6.6.2. As we stated, the crucial part of Algorithm 6 is the representation of the intermediate active areas (i.e. the regions R_i), since a good approximation of the exact active and inactive point sets is vital to remove as many inactive points as possible. In Section 7.3, an interval version of the elementary elimination procedure Step 4 of Algorithm 6 was implemented with rectangular approximations. We also noted that a more sophisticated method of Nurmela and Östergård [87] also exists in the literature. This latter method approximates the active and inactive regions by polygons. Although in a method working basically with multidimensional intervals the polygon representation is more difficult to implement, we found that this extra effort resulted in an outstanding improvement in computational efficiency compared to the method using rectangular approximations. Furthermore, since the branching step of the B&B frame algorithm generates rectangular splittings in a moderate way, our entire method actually combines the advantages of the two different approaches.

The method of [87] is based on the following lemma and theorem:

Lemma 8.1. [87]: *If a point p is at a distance less than $\tilde{f}_0$ from all the vertices of a polygon R, it is at a distance less than $\tilde{f}_0$ from all points of R.*

Theorem 8.2. [87]: *Assume that $p_1, \ldots, p_k$ are distinct points on the boundary of a polygon R_i, such that the line segments $\overline{p_l p_{l+1}}$ for $2 \leq l \leq k-2$ are edges of R_i, and that $\overline{p_1 p_2}$ and $\overline{p_{k-1} p_k}$ lie on the edges of R_i. If the points p_i, $1 \leq i \leq k$, are at a distance less than $\tilde{f}_0$ from all vertices of R_j $(i \neq j)$, then the points in the polygon formed by $p_1, p_2, \ldots, p_k$ are at a distance less than $\tilde{f}_0$ from all points of R_j.*

One can readily show that a set of points within a two-dimensional geometrical object having a distance of at least $\tilde{f}_0$ from all points of another object may be non-convex or non-connected. This fact, in addition to the rounding errors of the numerical implementation, requires special care to be taken for its mathematical correctness.

Let the polygon R_i be determined by the vertices $b_1, \ldots, b_s$, $s \geq 1$. By Theorem 8.2, the polygon formed by the convex hull of the points $p_1, \ldots, p_k$ contains only inactive points, i.e. it can be partly or fully eliminated.

Definition 8.3. *We call a polygon with consecutive vertices $p_1, \ldots, p_k$, $k \geq 3$, and with edges $\overline{p_1 p_2}, \ldots, \overline{p_{k-1} p_k}, \overline{p_k p_1}$ a* 'simple' polygon *if each pair of edges has at most one joined point as the joined endpoint of two consecutive edges.*

The following criterion will be stated for the new inactive point diminishing algorithm called Algorithm 7, which will be introduced later on in this section.

Invariance criterion: *during the execution of Algorithm 7, each R_i, $i = 1, \ldots, n$, will be either a point p_1, a line segment $\overline{p_1 p_2}$, or a 'simple' polygon with consecutive vertices $(p_1, \ldots, p_k)$, $k \geq 3$.*

In the rest of the chapter we will use the term 'polygon' for any figure which satisfies the above invariance criterion.

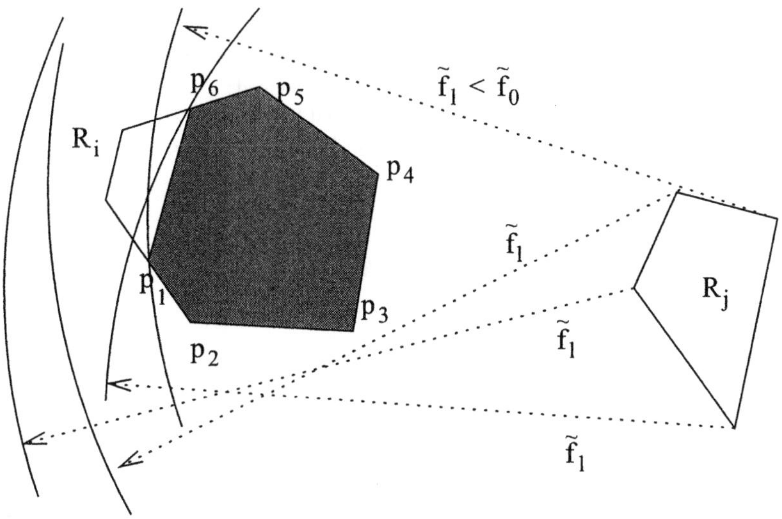

Fig. 8.1. A basic elimination procedure using polygons ($s = 6, k = 6$) with exact arithmetic. The shaded region of R_i can be regarded as the inactive region to be eliminated.

The pseudo code of the proposed interval version of an elementary polygon elimination step is given by Algorithm 7. That is, Algorithm 7 is a possible implementation of Step 4 of Algorithm 6. In Algorithm 7 we consider several cases that depend on s: $s = 1$ is handled in Steps 6 and 7, while $s = 2$ and $s \geq 3$ are considered in Steps 10 to 13, and Steps 14 to 20, respectively.

Note that in Algorithm 7 $p_0 = p_{k+1}$ may hold in Steps 15 and 16. In this case we construct R'_i without duplicating this point in the resulting polygon.

Assuming *exact computations*, one can easily prove that if the polygons $R_i, i = 1, \ldots, n$, are initialized as convex sets (as is the case in the current method—see Algorithm 6, Step 1), they remain convex during the steps of Algorithm 6, i.e. Step 20 of Algorithm 7 results in a convex polygon. But with finite precision machine arithmetic the points p_1 and p_k cannot be evaluated exactly (Algorithm 7, Steps 11, 16, and 17). In the method of Nurmela and Östergård the evaluated points p_1 and p_k are corrected by estimating the possible computation error, while in the present method proper rectangles as the guaranteed enclosures of p_1 and p_k are computed. However, both methods may result in concave, or even self-intersecting R_i polygons. To avoid the difficulty of representing and handling extremely irregular sets, we have to make some restrictions on the shape of the polygons. This was the reason for formulating the above invariance criterion.

The pseudo code of the proposed interval version of an elementary polygon elimination step is given by Algorithm 7. That is, Algorithm 7 is a possible implementation of Step 4 of Algorithm 6. In Algorithm 7 we consider several cases that depend on s: $s = 1$ is handled in Steps 6 and 7, while $s = 2$ and $s \geq 3$ are considered in Steps 10 to 13, and Steps 14 to 20, respectively.

Algorithm 7 : $Diminish\_ij()$ – an interval version of the polygon approximation

Inputs: – $R_i = R_i(b_1, b_2, \ldots, b_s)$: the polygon to be reduced,
– $R_j = R_j(a_1, a_2, \ldots, a_t)$: the polygon used for reducing R_i,
– $\tilde{f}_0$: a validated lower bound for the global maximum of Problem 6.1.

Output: – R_i': the remaining polygon of R_i.

1. **for** ($l := 1$ **to** s)
2. **if** (it is guaranteed that $d(b_l, a_m) < \tilde{f}_0,\ \forall m = 1, \ldots, t$) **then** mark b_l with a '$-$' flag;
3. **else** mark b_l with a '+' flag;
4. **endif**
5. **endfor**
6. **if** (all the b_l have '$-$') **then return** 'R_i' is empty';
7. **elsif** (all the b_l have '+') **then return** $R_i' := R_i$;
8. **endif**
9. Find a sequence of consecutive vertices b_l with '$-$', denoted by $p_2, \ldots, p_{k-1}$.
10. **if** ($s = 2$) *comment: R_i is a line segment*
11. Denote the node of R_i differing from p_2 by p_0;
12. Find an enclosure $P_1 \in \mathbb{I}^2$ of a point p_1 such that p_1 is on the line segment $\overline{p_0 p_2}$, and $d(p_1, a_m) < \tilde{f}_0,\ \forall m = 1, \ldots, t$.
13. Build R_i' from P_1, p_2;
14. **else** *comment: $s \geq 3$*
15. Denote the preceding node of p_2 in R_i by p_0;
16. Denote the succeeding node of p_{k-1} in R_i by p_{k+1};
17. Find an enclosure $P_1 \in \mathbb{I}^2$ of a point p_1 such that p_1 is on the line segment $\overline{p_0 p_2}$, and $d(p_1, a_m) < \tilde{f}_0,\ \forall m = 1, \ldots, t$.
18. Find an enclosure $P_k \in \mathbb{I}^2$ of a point p_k such that p_k is on the line segment $\overline{p_{k-1} p_{k+1}}$, and $d(p_k, a_m) < \tilde{f}_0,\ \forall m = 1, \ldots, t$.
19. Let $d_1 = p_{k+1}, \ldots, d_{s-k+2} = p_0$ be the consecutive vertices of R_i not chosen in Step 9;
20. Build R_i' from $P_1, P_k, d_1, \ldots, d_{s-k+2}$;
21. **endif**
22. **return** R_i';

We represent polygons commonly as a sequence of consecutive vertices, but we must assume that the coordinates of the vertices are machine numbers. (We start the elimination procedure with such polygons, see Algorithm 6, Step 1.) Each execution of Algorithm 7 results in either an empty set (Step 6), if we can *provide a guarantee* that each vertex of R_i is at a distance less than $\tilde{f}_0$ from R_j; or a polygon (Steps 7 or 20) which contains the polygon that would be obtained if exact arithmetic could be performed.

Remark 8.4. Notice that we need not assume any special properties of the sequence $p_2, \ldots, p_{k-1}$ of Step 9. Here the only requirement is that it consists of consecutive vertices of R_i marked with '$-$'. For example, if S is such a sequence, then any subsequence S' of consecutive vertices in S can also be considered.

Algorithm 8 : Step 12 and Step 17 of Algorithm 7

Inputs: – p_0, p_2: consecutive vertices of R_i. p_0 has the flag '+' and p_2 has the flag '–' (obtained by interval computations),
– $R_j = R_j(a_1, \dots, a_t)$: the reducing polygon,
– $F = \tilde{f}_l^2 \in \mathbb{I}$ such that $\tilde{f}_l < \tilde{f}_0$.
– $(X_i, Y_i) \in \mathbb{I}^2$: the ith initial active region.
Output: – an inclusion P_1 of an appropriate point p_1.

for $(m := 1$ **to** $t)$
  if $(\mathrm{ub}(D(p_0, a_m)) < \mathrm{lb}(F))$ **then** $C_m := p_0$;
  else $C_m := ComputeC(p_2, p_0, a_m, F)$;
  endif
endfor
$ind := \arg\min_{m=1}^{t} \mathrm{lb}(D(C_m, p_2))$;
$P_1 := C_{ind}$;
for $(m := 1$ **to** $t)$
  if $(\mathrm{ub}(D(C_{ind}, p_2)) \geq \mathrm{lb}(D(C_m, p_2)))$ **then** $P_1 := Comp\_Hull(P_1, C_m)$ **endif** ;
endfor
$P_1 := Intersection(P_1, (X_i, Y_i))$;
return P_1;

The first problem to be solved when implementing Algorithm 7 is that with the usual floating-point arithmetic one cannot reliably decide whether the distance between two points (represented by machine numbers) is less than a given machine number. Instead, we use interval arithmetic in the following way:

Marking the vertices of R_i by interval computations. Consider an arbitrary node $b_l(x_{b_l}, y_{b_l})$ of R_i, and denote the vertices of R_j by $a_1(x_{a_1}, y_{a_1}), \dots, a_t(x_{a_t}, y_{a_t}), t \geq 1$. Moreover, consider a machine number $\tilde{f}_l$ less than $\tilde{f}_0$. Such an $\tilde{f}_l$ can be determined by a direct downward rounding procedure offered by most interval packages. Now compute $D(b_l, a_m) := (x_{b_l} - x_{a_m})^2 + (y_{b_l} - y_{a_m})^2 \in \mathbb{I},\ m = 1, \dots, t$ by natural interval extension, considering each coordinate as a point interval and also compute $F := \tilde{f}_l^2 \in \mathbb{I}$ as an interval inclusion. If

$$\mathrm{ub}(D(b_l, a_m)) < \mathrm{lb}(F), \quad \forall m = 1, \dots, t,$$

then mark b_l with '–', otherwise mark b_l with '+'. Clearly, b_l receives the flag '–' only in cases where *it is guaranteed* that b_l is at a distance less than $\tilde{f}_l$, and thus less than $\tilde{f}_0$ from all the vertices of R_j. This guarantee has its cost however; the resultant set of nodes having the flag '–' may only be a subset of the set of nodes with flag '–' that would be obtained by assuming exact computations. However, in accordance with Remark 8.4 this fact has no effect on the correctness of Algorithm 7. It still runs as it should.

Computing inclusion rectangles for p_1 and for p_k. The second problem to be solved is that of finding a reliable alternative to the inaccurate floating-point computation of

p_1 and p_k. This is done in Steps 12, 17, and 18 of Algorithm 7. Here we will consider only the case of p_1 as a similar process can be introduced for p_k. Clearly, with exact computations, our aim would be *to find a point* p_1 *on the line segment* $\overline{p_0p_2}$ *that still can be eliminated, but which is as far from* p_2 *as possible*. Obviously, p_2 is a suitable choice for P_1 (since it has '$-$' flag). Then if problems (due to the overestimation) occur while computing P_1, Steps 12 and 17 can still return with $P_1 := p_2$. In the algorithm below we evaluate an inclusion (i.e. a rectangle) $P_1 \in \mathbb{I}^2$ of *an appropriate* p_1 point. Naturally we assume that p_1 is on the line determined by p_0 and p_2, and it is at a distance less than $\tilde{f}_0$ from R_j.

A procedure implementing Algorithm 8—i.e. Steps 12 and 17 of Algorithm 7 with exact computations—would work as follows (Figure 8.2a). First consider the half line H with endpoint p_2 and including p_0p_2. For each a_m compute a point c_m lying on H, where the distance of a_m and c_m is exactly $\tilde{f}_l$ (such c_m points must exist since p_2 has the '$-$' flag). Then it finds the c_m which is the closest to p_2 and sets p_1 to c_m. Figure 8.2a shows this method with $p_1 := c_3$. In contrast to the exact computation, the interval algorithm evaluates for each m either

(i) a two-dimensional point interval C_m on H which is not farther from p_2 than the exact c_m, or
(ii) a rectangle C_m containing the exact c_m.

Figure 8.2b shows the results for the interval version of Algorithm 8, where C_1 is determined by Step 2, while C_2 and C_3 are determined by Step 3.

The aim of the function call $ComputeC(p_2, p_0, a_m, F)$ is to produce an enclosure of the exact c_m when $\text{ub}(D(p_2, a_m)) < \text{lb}(F)$ (this holds since p_2 has a '$-$' flag), and additionally $\text{ub}(D(p_0, a_m)) \geq \text{lb}(F)$ (when the condition in Step 2 does not hold). In this case the exact c_m must lie on the line segment $\overline{p_0p_2}$, and $p_2 \neq c_m$. Notice that Step 3 of Algorithm 8 is always executed at least once, since p_0 has the flag '$+$'. Denoting the coordinates of the corresponding points in the usual way, we have to solve the following nonlinear system of equations for $c_m(x_{c_m}, y_{c_m})$ in a reliable manner:

$$(y_{c_m} - y_{p_0})(x_{p_2} - x_{p_0}) = (y_{p_2} - y_{p_0})(x_{c_m} - x_{p_0}),$$
$$(x_{c_m} - x_{a_m})^2 + (y_{c_m} - y_{a_m})^2 = \tilde{f}_l^2.$$

Here the first equation is equivalent to the statement that c_m lies on the line determined by p_0 and p_2 (when $p_0 \neq p_2$). The second equality tells us that the distance from c_m to a_m is $\tilde{f}_l$.

This system can be solved in the customary way by using interval computations (and handling the possibly different cases arising from the interval-valued discriminant). As the basic ideas are quite clear, the technical details for solving this system will not be presented here. The interval evaluations may result in significant overestimations (e.g. when one of the resulting rectangles $C_{m,1}, C_{m,2}$ contains p_2 or when the resulting boxes overlap), thus we accept the result of the solution procedure only in cases when $C_{m,1}$ and $C_{m,2}$ are disjunct, only one of them contains points from the half line H, and this solution does not contain p_2. (One can easily check the above criteria by comparing the bounds of $C_{m,1}, C_{m,2}, p_0$ and p_2.) In all the other cases the

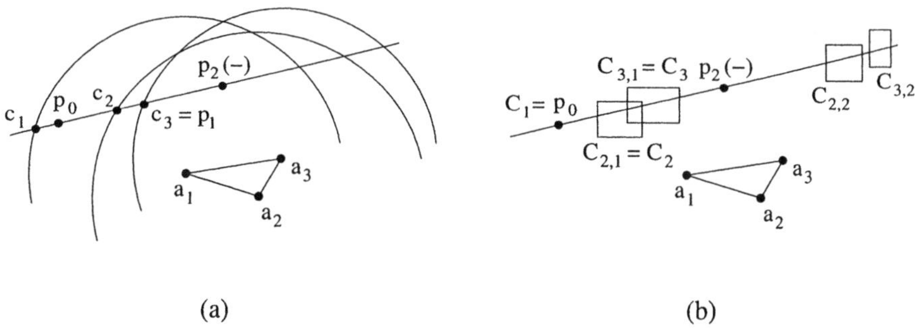

Fig. 8.2. Algorithm 8 with exact (left) and with interval (right) arithmetic.

algorithm returns $C_m := p_2$ as a 'safety solution'. In Figure 8.2b, both for $m = 2$ and for $m = 3$ the resulting rectangles $C_2 := C_{2,1}$ and $C_3 := C_{3,1}$ can be accepted.

Remark 8.5. In certain extreme cases the accepted C_m may not contain any points of the line segment $\overline{p_0p_2}$. Then C_m is set to p_0 and as a side effect we may obtain $P_1 = p_0$ as a result of Algorithm 8. This means that although p_0 has got the flag '+' by simple computations during the marking, it would also obtain the flag '$-$' after a more complicated process including e.g. ComputeC(). Obviously, this should happen very rarely in practice (and in fact it did not happen at all in our numerical studies). Noting this fact, in order to keep our whole method as simple as possible, we decided not to reverse the marking of p_0 in such cases.

Steps 6 to 9 of Algorithm 8 determine P_1 on the basis of the following principle. First consider all the possible sets of a number of t points where exactly one point is chosen from each C_m. Then for each set find the element closest to p_2 and assign a rectangular enclosure to the closest points. Such an enclosure is given by a componentwise union of several rectangles after we have executed Step 9 a couple of times. (The function call $Comp\_Hull(P_1, C_m)$ returns the componentwise hull of its two-dimensional interval arguments.) Steps 6 to 9 of Algorithm 8 implement the above procedure correctly due to the following. First, assume that there exists a point combination having its closest point to p_2 within $C_{m_1}, m_1 \neq ind$, where C_{m_1} is not added to P_1. Then $\text{ub}(D(C_{ind}, p_2)) < \text{lb}(D(C_{m_1}, p_2))$ by Step 9. This means that all the points in C_{ind} are closer to p_2 than any points in C_{m_1}, which contradicts the original assumption. In Figure 8.2b the value of ind can be set to 3, and additionally, if $\text{ub}(D(C_3, p_2)) \geq \text{lb}(D(C_2, p_2))$ and $\text{ub}(D(C_3, p_2)) < \text{lb}(D(C_1, p_2))$ hold, then P_1 is determined by the componentwise hull of C_2 and C_3.

Step 11 of Algorithm 8 does the rest of the work. Here since the ith initial active region is the rectangle (X_i, Y_i) (Algorithm 6, Step 1), the resulting polygon of the exact version of Algorithm 7 is included in (X_i, Y_i). Thus P_1 can be intersected by this rectangle. Note that the intersection of Step 11 is not empty: we know that $\overline{p_0p_2} \subseteq (X_i, Y_i)$ must contain a possible resulting point p_1.

Computing the resulting polygon R_i'. The only remaining problem to be solved for the interval version of the method of active areas is that of determining R_i' in some

---

Algorithm 9 : Step 20 of Algorithm 7

Inputs: – $d_1(=p_{k+1}),\ldots,d_{s-k+2}(=p_0)$: the non-empty set of consecutive vertices not selected in Step 9 of Algorithm 7,
– p_2, p_{k-1}: the first and the last element of the sequence of vertices selected in Step 9 of Algorithm 7,
– P_1, P_k: inclusions of p_1 and p_k, respectively,
– R_i: the polygon to be reduced.

Output: – R_i': the result polygon.

1. $K := Convex\_Hull(P_1, P_k)$;
2. **if** $(p_0 \neq p_{k+1})$ **then**
3.   **if** $((Separate(\{d_2,\ldots,d_{s-k+1}\},\{P_1,P_k\},\overline{p_0p_{k+1}})))$ **then**
4.   **if** $(K' := Convex\_Hull(p_0, p_{k+1}, K)$ is determined) **then**
5.   Let K' be denoted by $K'(p_0, e_1, \ldots, e_u, p_{k+1})$;
6. $R_i' := R_i'(e_1,\ldots,e_u,d_1,\ldots,d_{s-k+2})$;
7.   **return** R_i';
8.   **endif**
9.   **endif**
10.   **if** $((Separate(\{d_2,\ldots,d_{s-k+1}\},\{p_2,p_{k-1}\},\overline{p_0p_{k+1}})))$ **then**
11. $R_i' := R_i'(p_2, p_{k-1}, d_1, \ldots, d_{s-k+2})$;
12.   **return** R_i';
13.   **endif**
14. **else**   *comment:* $p_0 = p_{k+1}$
15.   **if** $(p_0 \notin K)$ **then**
16.   **if** $(R_i' := Convex\_Hull(p_0, K)$ is determined) **then** **return** R_i'; **endif**
17.   **endif**
18. **endif**
19. **return** R_i;

---

way, i.e. the implementation of Steps 13 and 20 of Algorithm 7. To do this we first consider the more complicated Step 20:

Implementation of Step 20 in Algorithm 7. We have the rectangles P_1 and P_k as the inclusions of the points p_1 and p_k, respectively, where p_1 and p_k are two suitable points (but not necessarily the same as for the exact algorithm) of the resulting polygon of Algorithm 7. For our interval implementation we will define the resulting polygon by vertices represented as machine numbers. This polygon includes *all the possible resulting polygons* where p_1 and p_k are chosen arbitrarily from P_1 and P_k, respectively. Our implementation is given by Algorithm 9. The essential workings of the algorithm is shown in Figure 8.3.

The key function call of Algorithm 9 is performed in Steps 3 and 10 and it is called $Separate()$. It has three parameters. The first one is a set of points, the second can be a set of points or a pair of rectangles, and the third is a line L defined by two points. $Separate()$ returns *true* only if it is guaranteed that all the elements of the

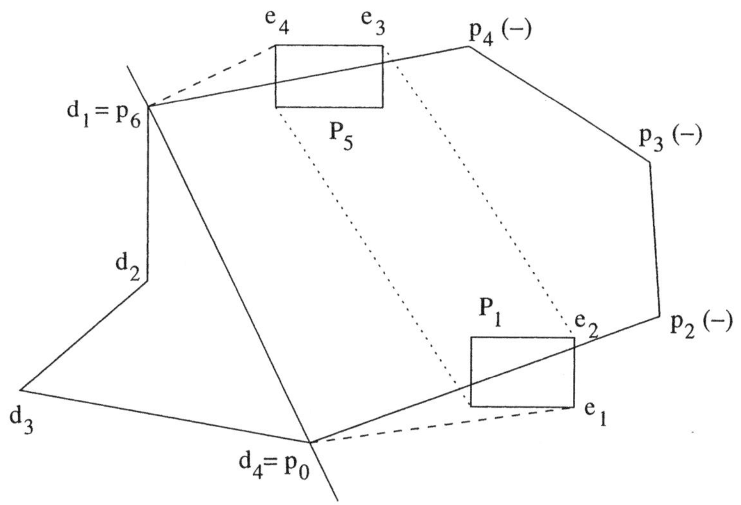

Fig. 8.3. An example of the application of Algorithm 9 for $s = 7, k = 5$, and $u = 4$. The resulting polygon is determined by the vertices $e_1, \ldots, e_4, d_1, \ldots, d_4$.

first parameter are located on the half plane determined by L and all the elements of the second parameter are located on the other half plane. We do not allow them to contact with line L. The above criterion seems to be a strict restriction, but it does help us to obtain polygons satisfying the required invariance criterion in a relatively easy way. During the numerical computations the value for $Separate()$ is returned true in most cases. Note that the first parameter set of $Separate()$ is allowed to be empty.

The other important function in Algorithm 9 is $ConvexHull()$, which returns the convex hull polygon of its argument. In general, this might be a difficult problem, especially with finite precision arithmetic. Nevertheless, since both P_1 and P_k are machine representable rectangles (or line segments or points in special cases) with horizontal and vertical bounds, in Step 1 $ConvexHull()$ can be easily determined. In Figure 8.3 the convex hull of P_1 and P_5 is determined by the two dotted line segments and the appropriate edges of P_1 and P_5.

The evaluation of the convex hull in Step 4 is a slightly harder task. Since it is called many times, we decided to code it ourselves for this special purpose instead of using standard interval tools. If $\{c_v\}$ denotes the set of vertices of K, we have to select the $e_1 := c_j$ point, for which the directed line segment $\overrightarrow{p_0c_j}$ has one of the following properties for all $\overrightarrow{p_0c_l}$, $l \neq j$: *either* $\overrightarrow{p_0c_j}$ is in the clockwise direction compared to $\overrightarrow{p_0c_l}$, *or* p_0, c_j and c_l are collinear and $\overline{p_0c_l} \subset \overline{p_0c_j}$.

In addition, we need another vertex of K (later denoted by e_u) for a similar process but here we need to consider p_{k+1} a the counter-clockwise orientation. To solve these subproblems, we invoke the basic element of a general method for generating convex hull sets (applied, of course, with interval arithmetic). Take, for example, $\overrightarrow{p_0c_j}$ and $\overrightarrow{p_0c_l}$ as two vectors in the *3-dimensional space* lying in the plane $z = 0$. Evaluate the interval enclosure of the *third component of the vector product* $(c_j - p_0) \times (c_l - p_0)$ using point interval arguments and interval arithmetic operations. Let us denote this enclosure by $P(\overrightarrow{p_0c_j}, \overrightarrow{p_0c_l}) \in \mathbb{I}$. Now,

(i) if $\mathrm{ub}(P(\overrightarrow{p_0c_j}, \overrightarrow{p_0c_l})) < 0$, then it is certain that $\overrightarrow{p_0c_j}$ is in a counter clockwise direction compared to $\overrightarrow{p_0c_l}$;
(ii) if $\mathrm{lb}(P(\overrightarrow{p_0c_j}, \overrightarrow{p_0c_l})) > 0$, then it is guaranteed that $\overrightarrow{p_0c_j}$ has in a clockwise direction compared to $\overrightarrow{p_0c_l}$;
(iii) if $P(\overrightarrow{p_0c_j}, \overrightarrow{p_0c_l}) = [0, 0]$, then it is ceratin that p_0, c_j and c_l are collinear. If this is the case but $\overline{p_0c_l} \subset \overline{p_0c_j}$ cannot be proved, then we declare that the convex hull is undetermined;
(iv) in all the other cases we declare that the convex hull is undetermined.

If the condition in Step 2 of Algorithm 9 holds, then the line segment $\overline{p_0p_{k+1}}$ is defined. If the condition of Step 3 is false or K' cannot be determined, then we try to produce an output polygon in Steps 10 to 12 using p_2 and p_{k-1} instead of P_1 and P_k.

If $p_0 = p_{k+1}$, then L is undefined. In this case (Steps 15 to 17), the set $\{d_2, \ldots, d_{s-k+1}\}$ should necessarily be empty (due to the invariance property of R_i). In Step 15 we test a property similar to the function $Separate()$: if p_0 is guaranteed to be outside of K, then try to determine the convex hull of p_0 and K by a similar process introduced for Step 4.

Finally, if the creation of R_i' was not possible, in Step 19 we return to the original R_i active region.

The implementation of Step 13 in Algorithm 7: Notice that this step can be realized as a special case of Algorithm 9: we have to determine the convex hull of a point p_2 and the inclusion rectangle P_1. Obviously, Steps 15 to 17 of Algorithm 9 implement this correctly with $K := P_1$.

As the final step, we prove the correctness of the interval implementation of Algorithm 7:

Theorem 8.6. *The interval implementation of Algorithm 7 eliminates only those points from R_i, which are guaranteed to be at a distance less than $\tilde{f}_0$ from all points of R_j.*

PROOF. First, we show that the output polygons satisfy the required invariance property. The initial R_i (Algorithm 6, Step 1) is either a point or a line segment or a rectangle. Assume that the input polygon R_i of Algorithm 7 satisfies the invariance property. If R_i is a point then the output of Algorithm 7 is either a point or an empty polygon (Algorithm 7, Steps 6 and 7). When R_i has several nodes, one can easily see that the separation tests and the properties of obtaining convex hulls described in Algorithm 9 guarantee that the edges of the output polygon can only touch in the required way. (Recall that we allow concave shapes as resulting polygons.)

Let $R_i = R_i(b_1, \ldots, b_s)$ be the input polygon of Algorithm 7 (satisfying the invariance criterion) and let R_i' be the output polygon produced by the interval version of Algorithm 7. Consider the following cases:

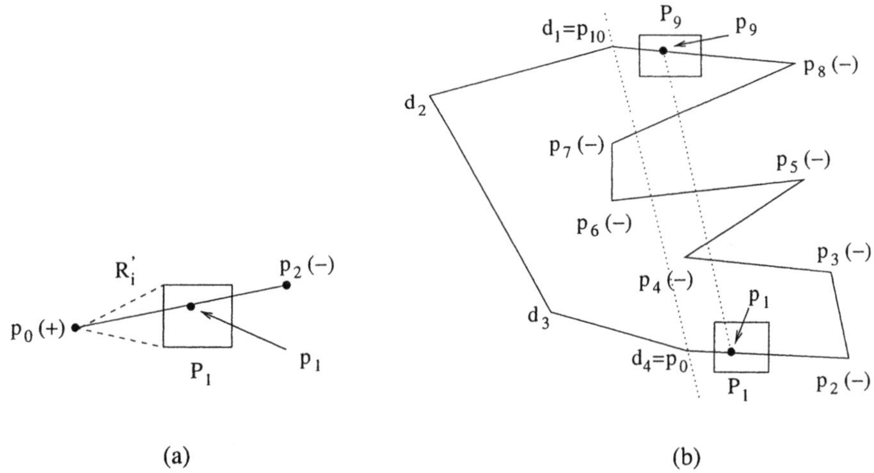

Fig. 8.4. An illustration for the proof of the correctness of Theorem 8.6. Case (3b) for $s = 2$ (left) and case (3c) for $s \geq 3$ with $s = 11,\ k = 9$ (right).

(1) R_i' is empty. This result can be achieved only in Step 6 of Algorithm 7, i.e. when all nodes of R_i get the flag '$-$'. Due to the reliable marking of the vertices, this statement holds true only if the exact computations would also give the '$-$' flag for all vertices, when the whole polygon could be deleted by Theorem 8.2.

(2) $R_i' = R_i$. The interval algorithm variant returns this result either if all vertices of R_i are labeled by a '+' (Algorithm 7, Step 7), or if a 'failsafe solution' is produced due to the overestimation or due to some technical difficulties (in Algorithms 8 and 9). In this case we did not remove any possible inactive points from R_i.

(3) $R_i' \neq R_i$ and R_i' is not empty. We investigate this case for different s values:

(3a) The case $s = 1$ cannot occur here since it is covered by the parts (1) and (2) of the proof.

(3b) If $s = 2$, we perform Steps 15 to 17 of Algorithm 9 instead of the original Step 13 of Algorithm 7, as discussed above. In these steps we find that $p_0 \notin P_1$ and then successfully determine R_i' as the convex hull of a point p_0 and a rectangle P_1 (see Figure 8.4a). Here P_1 is either a rectangle (with $p_0, p_2 \notin P_1$) or $P_1 \equiv p_2$ (a 'failsafe solution') produced by Algorithm 8, and it is ensured, that P_1 contains a point p_1 which is at a distance less than $\tilde{f}_0$ from R_j. Hence only the line segment $\overline{p_0 p_1}$ must belong to the remaining region. From the definition of the convex hull this obviously holds true.

(3c) If $s \geq 3$, we perform Algorithm 9 as Step 20 of Algorithm 7. This case is shown in Figure 8.4b. Consider the input polygon R_i with a non-empty set of consecutive vertices $p_2, \ldots, p_{k-1}$ labeled by '$-$'. The remaining non-empty set of nodes is denoted by $d_1 = p_{k+1}, d_2, \ldots, d_{s-k+2} = p_0$. Here $p_0 = p_{k+1}$ is also possible (see Steps 14 to 16 of Algorithm 9).

With our assumption, P_1 and P_k are successfully generated by Algorithm 8. Namely, P_1 is either a rectangle (with $p_2 \notin P_1$) or $P_1 \equiv p_2$, and it is guaranteed that

P_1 contains a point p_1 which is at a distance less than $\tilde{f}_0$ from R_j. Obviously, similar statements apply for P_k (with p_{k-1}).

Moreover, the required separation properties tested in Algorithm 9 are satisfied (implying that $p_0 \notin P_1$ and $p_{k+1} \notin P_k$, respectively) and a reliable convex hull of the set $\{p_0, p_{k+1}, P_1, P_k\}$ is determined.

Denote by P_+ the polygon determined by its consecutive vertices $p_1, p_k, d_1, \ldots, d_{s-k+2}$. Similarly, denote by P_- the polygon determined by the consecutive vertices $p_1, p_2, \ldots, p_k$. First of all, P_+ satisfies the invariance property (since p_1 and p_k are separated from the half plane containing all the d_j points), and from the construction of Algorithm 9 $P_+ \subseteq R_i'$ holds. Secondly, consider a point $p \in R_i$, $p \notin P_+$. Note that P_- can be a self-intersecting polygon, but in P_- only $\overline{p_1p_k}$ can cross some other edges (because of the invariance property of R_i). Thus p must be located in one of the pieces determined by the possible intersection points and the vertices of P_-. In any case, $p \in Conv(P_-)$ must hold, where $Conv(P_-)$ denotes the convex hull of P_-. Overall we obtain the relation

$$R_i \subseteq P_+ \cup Conv(P_-) \subseteq R_i' \cup Conv(P_-). \tag{8.1}$$

Assume now that a point $p \in R_i$ is eliminated by the interval implementation of Algorithm 7, i.e. $p \notin R_i'$. Then $p \in Conv(P_-)$ by (8.1). Recall that $p_2, \ldots, p_{k-1}$ have the flag '$-$', and that p_1 and p_k can also get '$-$' by definition. By Theorem 8.2 this means that $Conv(P_-)$ can fully be eliminated (assuming exact computations). In other words, p is eliminated by the interval method in the correct way. □

Corollary 8.7. *Using the above interval implementation for the elementary reduction step of the method of active regions, Algorithm 6 deletes only those (x, y) feasible points for which $f_n(x, y) < \tilde{f}$ holds.*

PROOF. Consider the remaining regions R_k, $k = 1, \ldots, n$ at any time while executing Algorithm 6, and assume that $(x_i', y_i') \in R_i$ is deleted by Algorithm 7, i.e. (x_i', y_i') is at a distance less than $\tilde{f}_0$ from an $R_j, j \neq i$ region. This means that we delete all the feasible solutions (x, y), for which $(x_i, y_i) = (x_i', y_i')$ and $(x_k, y_k) \in R_k$, $\forall k = 1, \ldots, n$, $k \neq i$ holds. By Theorem 8.6, the distance between (x_i, y_i) and (x_j, y_j) is less than $\tilde{f}_0$, hence from (6.8) the squared distance between them is guaranteed to be less than $\tilde{f}$. Consequently, $f_n(x, y) \leq (x_i - x_j)^2 + (y_i - y_j)^2 < \tilde{f}$, which completes the proof. □

8.2 A new technique for eliminating tile combinations

In Section 7.4.2 we mentioned the difficulties encountered when trying to find globally optimal solutions for point or circle packing problems. We discussed a general idea called *tiling*, which enabled us to determine the global solutions for up to $n = 27$

circles in a numerical way. However, we pointed out that due to the exponentially increasing number of tile combinations, the sequential investigation of all tile combinations can only be carried out using a few decades of CPU time when $n = 28$ or slightly more than that. For instance, for $n = 28$ the 7×6 tiling needs $\binom{42}{28} \approx 5.3 \cdot 10^{10}$ combinations to be checked. As in earlier studies, symmetry properties may help to reduce this number, but even with these tricks we would still have to deal with more than $1.3 \cdot 10^{10}$ cases.

The main idea that helps us overcome this problem is the following: a large part of the combinations may be eliminated in a relatively easy way by considering only local relations within the combinations. In detail, if one can discover patterns of tiles which cannot contain components of an optimal solution, then several tile combinations (i.e. higher-dimensional subproblems) containing any of these patterns can readily be discarded. Let us denote a point arrangement problem instance by $P(n, X_1, \dots, X_n, Y_1, \dots, Y_n)$ where n is the number of points to be arranged ($2n$ is the dimensionality of the problem), $X_i, Y_i \in \mathbb{I}$, $i = 1, \dots n$ are the components of the starting box, and the objective function is given by (6.3). The theorem below shows how to apply a result obtained on a $2m$-dimensional packing problem for a higher-dimensional problem with $2n$ variables, where $n \geq m \geq 2$.

Theorem 8.8. *Let $n \geq m \geq 2$ be integers and let*

$$P_m = P(m, Z_1, \dots, Z_m, W_1, \dots, W_m) = P(m, (Z, W)), \text{ and}$$
$$P_n = P(n, X_1, \dots, X_n, Y_1, \dots, Y_n) = P(n, (X, Y))$$

be point arrangement problem instances ($X_i, Y_i, Z_i, W_i \in \mathbb{I}$; $X_i, Y_i, Z_i, W_i \subseteq [0, 1]$). Run Algorithm 5 on P_m using a hypothetic $\tilde{f}$ cutoff value in the accelerating devices, skip Step 7, and stop after an arbitrary, preset number of iteration steps. Denote by $(Z_1', \dots, Z_m', W_1', \dots, W_m') := (Z', W')$ the componentwise hull of all the elements placed on the WorkList and on the ResultList. Assume that there exists an invertible, distance-preserving geometrical transformation φ with $\varphi(Z_i) = X_i$ and $\varphi(W_i) = Y_i$, $\forall i = 1, \dots, m$. Then for each point arrangement $(x, y) \in \mathbb{R}^{2n}$ satisfying $(x, y) \in (X, Y)$ and $f_n(x, y) \geq \tilde{f}$, the statement

$$(x, y) \in (\varphi(Z_1'), \dots, \varphi(Z_m'), X_{m+1}, \dots, X_n,$$
$$\varphi(W_1'), \dots, \varphi(W_m'), Y_{m+1}, \dots, Y_n) := (X', Y')$$

also holds.

PROOF. Perform an indirect proof and assume that a feasible solution $(x, y) \in \mathbb{R}^{2n}$ of P_n with $f_n(x, y) \geq \tilde{f}$ is discarded when modifying the starting box of P_n, i.e. $(x, y) \in (X, Y)$, but $(x, y) \notin (X', Y')$. Additionally, let $(z, w) = (\varphi^{-1}(x_1), \dots, \varphi^{-1}(x_m), \varphi^{-1}(y_1), \dots, \varphi^{-1}(y_m)) \in \mathbb{R}^{2m}$. Notice that $(x, y) \notin (X', Y')$ implies $(z, w) \notin (Z', W')$ by the indirect assumption, and, moreover that,

$$\tilde{f} \leq f_n(x, y) \leq f_m(x_1, \dots, x_m, y_1, \dots, y_m) = f_m(z, w). \tag{8.2}$$

The second inequality of (8.2) follows from equation (6.3) and the equation in (8.2) is implied by the distance-preserving property of φ^{-1}. Consequently, Algorithm 5

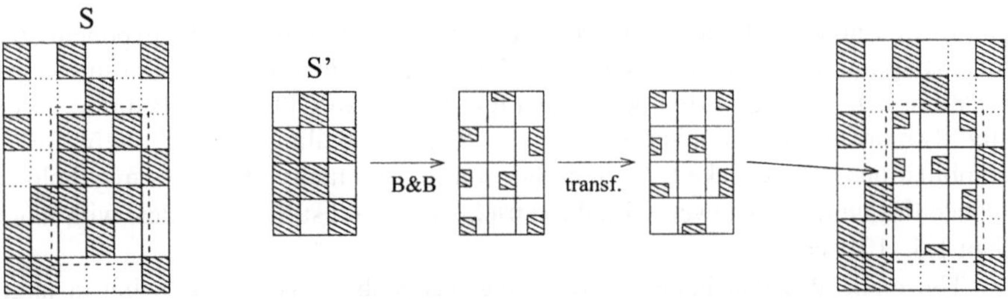

Fig. 8.5. The idea behind Theorem 8.8. The particular symmetry transformation applied is a reflection to the horizontal bisector of the region enclosing S'.

must have deleted the feasible solution (z, w) of P_m for which $\tilde{f} \leq f_m(z, w)$ holds. But this is a contradiction, since all the accelerating devices of Algorithm 5 delete only those feasible solutions for which the objective function value is less than $\tilde{f}$. This completes the proof. □

The meaning of Theorem 8.8 is the following: we assume that we are able to reduce some search regions on a tile set S'. When processing a higher-dimensional subproblem on a tile set S containing the image of S', it is enough to consider *the image of those of the remaining regions of* S' as the particular components of S. The idea is demonstrated in Figure 8.5. In the figure the regions of S and S' are denoted by shading, and the applied φ transformation is a reflection to the horizontal bisector of the rectangular enclosure of S'. Usually we apply the theorem in the special case when φ is the identity, i.e. when the tile set of the higher-dimensional problem is the superset of the tile set of the smaller problem.

As a consequence of the theorem, if it is proved that S' cannot contain point packings having at least the $\tilde{f}$ function value, then all the higher-dimensional problems with tile set S, $S' \subseteq S$ could be eliminated at once (when using the same $\tilde{f}$). The latter fact is formally stated below with the notation of Theorem 8.8:

Corollary 8.9. *Let φ be the identity transformation and assume that Algorithm 5 stops with an empty WorkList and with an empty ResultList, i.e. the whole search region $(Z, W) = (Z_1, \ldots, Z_m, W_1, \ldots, W_m) = (X_1, \ldots, X_m, Y_1, \ldots, Y_m)$ is eliminated by the accelerating devices using (the same) $\tilde{f}$. Then (X, Y) does not contain any $(x, y) \in \mathbb{R}^{2n}$ vectors for which $f_n(x, y) \geq \tilde{f}$ holds.*

Remark 8.10. In order to apply Theorem 8.8 correctly, we have to take into account the following important facts:

1. $w(Z_i) = w(X_i)$ and $w(W_i) = w(Y_i)$, $\forall i = 1, \ldots, m$ should be provided. This assumption does not necessarily hold if one tries to split the unit square in a regular way using floating-point numbers. The solution is the enlargement of the unit square in such a way that the resulting bounds of the tiles are exactly representable machine numbers (e.g. small integers).

2. We use shifting and rotating operators as φ, but even in these simple cases we have to apply the transformations in an interval way to obtain a guaranteed enclosure of the transformed objects.

8.2.1 Basic algorithms for the optimality proofs

Our optimality proofs are based on Theorem 8.8 and have the following basic idea: first we find feasible patterns of tiles and remaining areas on some small subsets of the whole set of tiles and then we process bigger and bigger subsets while using the results of the previous steps. Thus, our method consists of several phases, and each phase depends on the result of the previous phases. (On the computer, the multiphase method is controlled by short command scripts related to each phase.)

Obviously, there are several ways of 'growing' the considered search regions until we reach the enclosures of the global maximizers as our final goal. We introduce two basic algorithms executed in our computer-aided proofs. We discuss them in general, i.e. for an arbitrary n and for a regular $k \times l$ splitting (with k rows and l columns) of the square. We assume that the columns are numbered by $1, \ldots, l$ from left to right. In order to discuss the algorithms in detail, we need to introduce some new notation and abbreviations.

At the beginning of each phase we determine some sets of tile combinations consisting either of the remaining areas of the previous phases (or its transformations) or of full tiles. These sets will be denoted by $S^{m}_{s..f}$, where $s \leq f$, $s, f \in \{1, \ldots, l\}$, $m \in \{0, 1, \ldots, n\}$, symbolizing that a number of m remaining or full tile regions are considered and the regions are chosen from columns $s, s+1, \ldots, f$. (In the example of $n = 29$, we first generate $S^{15}_{1..3}$, that is, all the combinations of 15 tiles where the tiles are chosen from the leftmost three columns of the square.) Moreover, for a set $M = \{m_1, \ldots, m_k\} \subseteq \{0, 1, \ldots, n\}$, let $S^{M}_{s..f} = \cup_{i=1}^{k} S^{m_i}_{s..f}$. The elements of the sets $S^{m}_{s..f}$ are processed sequentially by the optimization algorithm using $\tilde{f}$. With the exception of the final phase, we halt the optimization algorithm after a certain number of iterations. As a result, some combinations can be fully or partly eliminated. The resulting new sets will be denoted by $\bar{S}^{m}_{s..f}$. The sets $\bar{S}^{m}_{s..f}$ form the input components of the subsequent phases. We will use the notation ${}^{m_1}_{s_1..f_1}S^{m}_{s..f}$, $1 \leq s \leq s_1 \leq f_1 \leq f \leq l$, $0 \leq m_1 \leq m \leq n$ to represent the subset of $S^{m}_{s..f}$ in which each element contains exactly m_1 tile regions from columns $s_1, \ldots, f_1$. Note that the elements of the sets $S^{m}_{s..f}$, $\bar{S}^{m}_{s..f}$, ${}^{m_1}_{s_1..f_1}S^{m}_{s..f}$, and ${}^{m_1}_{s_1..f_1}\bar{S}^{m}_{s..f}$ are in general only *enclosures* of the corresponding exact remaining areas.

Remark 8.11. In each phase of our optimality proof we accelerate Algorithm 5 by calling the method of active areas immediately to the initial search region. In many cases this results in the elimination of the whole search box before doing any bisection. Obviously, this modification has no effect on the correctness of Theorem 8.8.

Algorithm 10 takes the set $\bar{S}^{m}_{1..c}$ for a given pair (c, m), i.e. all the current remaining areas where m tiles are chosen from the first c columns and produces a new set

Algorithm 10 : *Grow* – add one column of tiles to the elements of a set of tile regions

Inputs: – n: the number of points to be packed,
– k, l: the square is divided into k rows and l columns regularly,
– c, $1 \le c \le l$: a column index,
– $\bar{S}^t_{1..c}$ for all $0 \le t \le ck$,
– $\bar{S}^t_{c+1..l}$ for all $0 \le t \le (l-c)k$ (if available),
– m, $1 \le m \le ck$: the elements of $\bar{S}^m_{1..c}$ will be extended by adding one column,
– i: the number of tiles to be added from column $c+1$.

Output: – ${}^{m}_{1..c}S^{m+i}_{1..(c+1)}$: the output set.

1. Initialization: ${}^{m}_{1..c}S^{m+i}_{1..(c+1)} := \emptyset$;
2. Generate $S^i_{(c+1)..(c+1)}$ from full tiles;
3. **forall** $(A \in \bar{S}^m_{1..c})$
4. Let j be the number of regions in columns $2, \ldots, c$ of A;
5. **forall** $(B \in S^i_{(c+1)..(c+1)})$
6. **if** ($\bar{S}^{n-m}_{c+1..l}$ is available and there exists an element of it having the same tile pattern as B in column $c+1$) **then**
7. Let A' be the concatenation of A and B;
8. Let P be the set of tile positions in columns $2, \ldots, (c+1)$ of A';
9. **if** (a $C \in \bar{S}^{i+j}_{2..(c+1)}$ exists with tile positions P) **then**
10. Intersect A' and C on positions P resulting A'';
11. **if** (none of the components of A'' is empty) **then**
12. Add A'' to ${}^{m}_{1..c}S^{m+i}_{1..(c+1)}$;
13. **endif**
14. **endif**
15. **endif**
16. **endfor**
17. **endfor**

${}^{m}_{1..c}S^{m+i}_{1..(c+1)}$. The procedure takes all the combinations $A \in \bar{S}^m_{1..c}$ (Step 3) and adds a suitable (see Remark 8.12 below) number i of new tiles to them in every possible way. With the aid of the sets $\bar{S}^{n-m}_{c+1..l}$, in Step 6 we test whether pattern B can be valid in column $(c+1)$. If this is the case, each result combination A' (Step 7) can be tested further since we have additional information on the possible remaining areas in columns $2, \ldots, (c+1)$. Namely, $\bar{S}^{i+j}_{2..(c+1)}$ can be generated from the given $\bar{S}^{i+j}_{1..c}$ by applying Theorem 8.8 (performing a shifting transformation). Should there is no remaining combination with a pattern P in $\bar{S}^{i+j}_{2..(c+1)}$ (tested in Step 9), we can immediately ignore A' by Corollary 8.9. Otherwise we intersect A' with $C \in \bar{S}^{i+j}_{2..(c+1)}$ on the tile positions corresponding to P (Step 10), and apply Theorem 8.8 again to the result A''. We store A'' in the output set only if none of its components are empty (Steps 11 and 12).

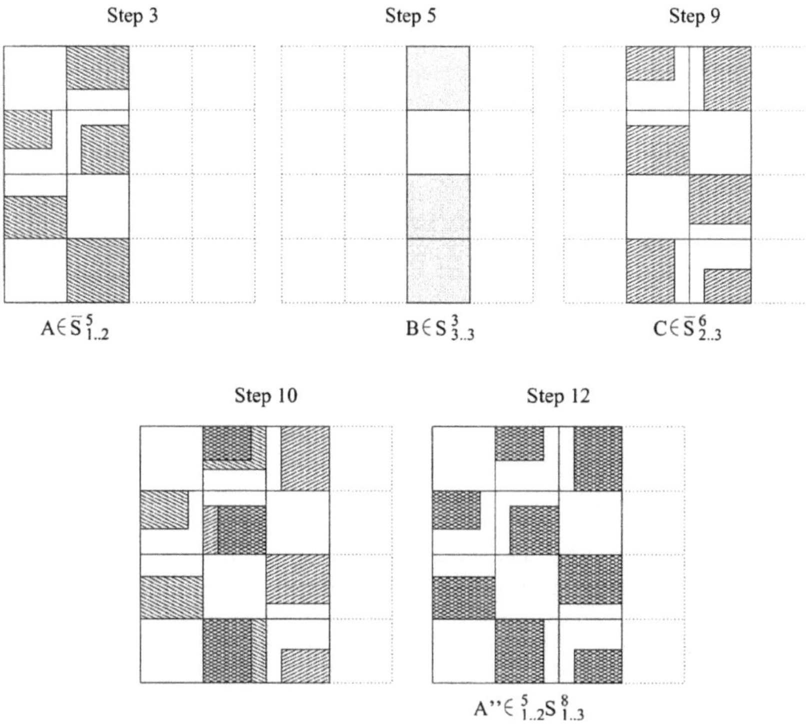

Fig. 8.6. An example of adding a new column with $k = l = 4,\ c = 2,\ m = 5,\ i = 3,\ j = 3$.

Figure 8.6 shows how the algorithm works with $k = l = 4,\ c = 2,\ m = 5,\ i = 3$, and with $j = 3$ in the main loop. In each part of the figure solid lines denote the boundaries of the combinations in question.

Remark 8.12. In practice, Algorithm 10 is performed on some relevant m values. In order to cover all the valid tile combinations on the first $c + 1$ columns, we have to compute a lower m_l and an upper m_u bound on the number of tiles $m+i$ of the output, i.e. to associate a set of i values with each m. A possible choice for a lower bound is m; however, we may increase this number if we have a kind of complementary information: assume that one can guarantee that all the point arrangements $x \in \mathbb{R}^{2n}$, $f_n(x) \geq \tilde{f}$ contain at most t points in columns $(c + 2), \ldots, l$. In other words, we know that $S^{t_0}_{c+2..l} = \emptyset$ (or $\bar{S}^{t_0}_{c+2..l} = \emptyset$) for each $t_0 > t$. Then m_l can be set up to $\max(m, n - t)$. An upper bound on $m + i$ can be found by using $m_u := \min(m + k, n)$. Summarizing these results, Algorithm 10 must be performed for all i with $\max(0, n - t - m) \leq i \leq \min(k, n - m)$ for each m.

Algorithm 11 joins the elements of two sets of tile combinations in a pairwise fashion. This algorithm also needs a given pair (c, m) as input data. Here it is assumed that the possible remaining regions are known when locating m points in the first c columns. Additionally, we have all the possibilities for adding some points in the

Algorithm 11 : *Join* – join the remaining two regions of tile sets

Inputs: – n: the number of points needed for packing,
– k, l: the square is divided into k rows and l columns in a regular way,
– c, $1 \le c \le l-1$: a column index,
– m, $0 \le m \le ck$,
– ${}^{m}_{1..c}\bar{S}^{m_1}_{1..(c+1)}$ for all $m_1, m \le m_1 \le n$,
– ${}^{n-m}_{(c+1)..l}\bar{S}^{m_2}_{c..l}$ for all $m_2, n-m \le m_2 \le n$.
Output: – ${}^{m}_{1..c}S^{n}_{1..l}$: the output set.

1. Initialization: ${}^{m}_{1..c}S^{n}_{1..l} := \emptyset$;
2. **for** $(m_1 := m$ **to** $n)$
3.   **forall** $(A \in {}^{m}_{1..c}\bar{S}^{m_1}_{1..(c+1)})$
4.   Let j be the number of regions in column c of A;
5.   Let P be the set of tile positions in columns c and $(c+1)$ of A;
6.   **forall** $(C \in {}^{n-m}_{(c+1)..l}\bar{S}^{n-m+j}_{c..l})$
7.   Let P' be the set of tile positions in columns c and $(c+1)$ of C;
8.   **if** $(P = P')$ **then**
9.   Intersect A and C at positions P resulting in A';
10.   **if** (none of the components of A' are empty) **then**
11.   Add A' to ${}^{m}_{1..c}S^{n}_{1..l}$;
12.   **endif**
13.   **endif**
14.   **endfor**
15.   **endfor**
16. **endfor**

$(c+1)$th column for each element of $\bar{S}^{m}_{1..c}$. Thus we know the sets ${}^{m}_{1..c}\bar{S}^{m_1}_{1..(c+1)}$, by Algorithm 10, say. Similarly, the sets ${}^{n-m}_{(c+1)..l}\bar{S}^{m_2}_{c..l}$ are also assumed to be known in advance. (In practice, the latter sets are evaluated from ${}^{n-m}_{1..(l-c)}\bar{S}^{m_2}_{1..(l-c+1)}$ by applying a rotation of 180 degrees at the midpoint of the search square for each element, since we 'grow' tile combinations from left to right, starting from column 1.) Consequently, the input sets can result in intersected tile positions from columns c and $c+1$.

Algorithm 11 works as follows. First we take all the elements A from ${}^{m}_{1..c}\bar{S}^{m_1}_{1..(c+1)}$ (Step 3). Then, for each A, we determine *all* those combinations $C \in {}^{n-m}_{(c+1)..l}\bar{S}^{n-m+j}_{c..l}$ that have the same tile positions in columns c and $c+1$ as A (Steps 5 – 8). If the condition in Step 8 holds, then we join A and C by Theorem 8.8 (Step 9). And finally, similar to Algorithm 10, we store the result only if all of its components are non-empty (Steps 10 and 11). The output set contains all those remaining areas of the considered packing problem for which exactly m tiles are chosen from the leftmost c columns. Figure 8.7 shows an example of Algorithm 11 with $n = 11$, $k = l = 4$, $c = 2$, $m = 6$ and for $m_1 = 8$, $j = 3$.

Notice that Algorithm 11 can be generalized on the basis of the number i of intersecting columns. Similar representations with $i = 0, \ldots, l-1$ are possible as

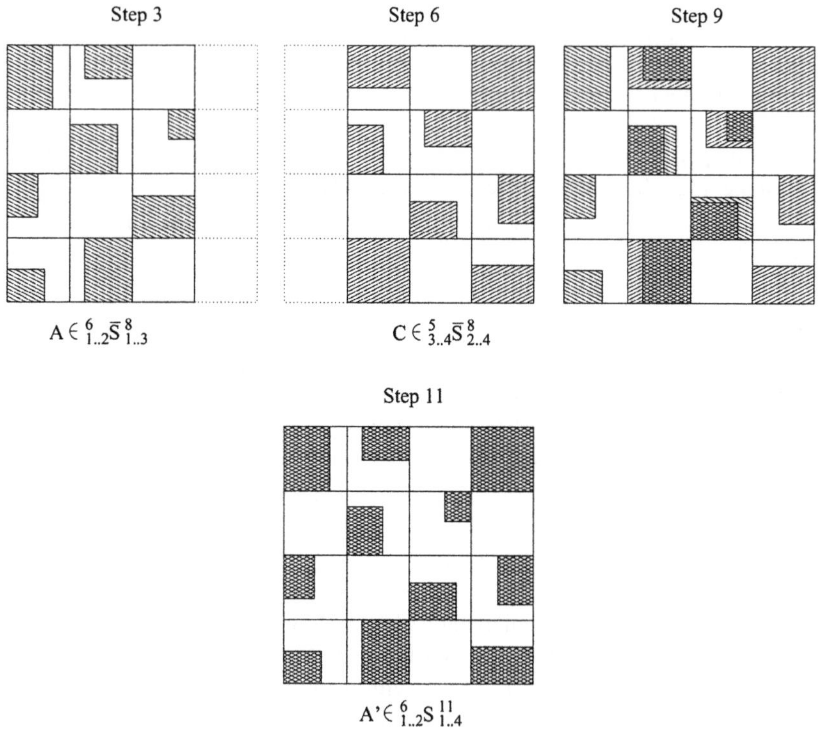

Fig. 8.7. An example of joining the remaining areas for $n = 11,\ k = l = 4,\ c = 2,\ m = 6$, and for $m_1 = 8,\ j = 3$.

well. Here only $i = 2$ is considered since it proved to be quite sufficient for the problem instances discussed earlier. However, in general it may be worth applying it to many different i values.

8.3 An improved method for handling free points

In Section 7.2 we stressed the importance of recognizing the occurrences of *free points* in point arrangement configurations. Below we introduce an improved version of the first such method proposed in Section 7.2.

In practice, the enclosure of a set containing only free points appears as a region that is much bigger than the others (and it is probably subdivided unnecessarily), but it is quite useless as a reducing region. It is apparent that one should consider only a single canonical value rather than of such a point set (just like the method of Section 7.2). The following technique shows how to apply a simple criterion that allows one to temporarily replace some parts of the search region by single points *without losing any optimal solutions*.

1. Let $(X_1,\dots,X_n,Y_1,\dots,Y_n) = (X,Y) \in \mathbb{I}^{2n}$ enclose all the remaining boxes (i.e. all the candidates stored either in the WorkList or in the ResultList) after a certain number of iteration loops when executing Algorithm 5 (or its accelerated version—see Remark 8.11). We assume that either the WorkList or the ResultList is not empty. Moreover, let $\tilde{f}$ be the current cutoff value used in the algorithm.
2. Assume that there exist machine representable points $p_{k_1},\dots,p_{k_t},\ p_{k_s} \in (X_{k_s},Y_{k_s})$, $s \in \{1,\dots,t\}$, in t different components of (X,Y), such that for each k_s,
$$\mathrm{lb}(D(p_{k_s},(X_j,Y_j))) > \mathrm{ub}(F_n(X,Y)) \geq \tilde{f} \tag{8.3}$$
holds for all $j \in \{1,\dots,n\}$, $j \neq k_s$. Let K denote the set of indices $\{k_1,\dots,k_t\}$.
3. Replace the components (X_i,Y_i) of (X,Y) with the point intervals p_i for each $i \in K$. Run Algorithm 5 on the resulting (X',Y') box, ignoring the improvement step of $\tilde{f}$, i.e. using the earlier found best lower bound, and terminate it after a certain number of iteration steps.
4. Let $(X'',Y'') \in \mathbb{I}^{2n}$ enclose all the remaining boxes. The components of the *output box* of the procedure are then given by (X_i,Y_i) for $i \in K$, and by (X''_j,Y''_j) for each $j \notin K$; i.e. the latter components can be reduced relative to (X_j,Y_j).

The idea behind the proposed method is the following: assume that optimal point packings exist in (X,Y). Then clearly $\mathrm{ub}(F_n(X,Y)) \geq f^*$ holds. Consider a set of points $\{p_j \mid p_j \in (X_j,Y_j), j \notin K\}$ with a minimal pairwise distance of f^*. Notice that by (8.3) this set can be expanded with $\{p_i, i \in K\}$ to obtain an optimal packing. Moreover, for each $i \in K$, $d(p_i,p_j) > \mathrm{ub}(F_n(X,Y)) \geq f^*$ occurs for all $j \in \{1,\dots,n\}$, $j \neq i$, which implies the existence of an $\varepsilon > 0$ such that $d(p'_i,p_j) > f^*$, $\forall p'_i \in N_\varepsilon(p_i)$. In other words, each p_i is a free point of each such packing.

Theorem 8.13. *The above procedure for shrinking the remaining regions is correct in the sense that all the optimal packings in (X,Y) are also present in the output box of the procedure.*

PROOF. Let us use the abbreviations $Z := (X,Y)$ and $Z' := (X',Y')$. Consider a box $V' \subseteq Z'$ eliminated via Algorithm 5 using $\tilde{f}$ in Step 3 of the above method. Recall that $V'_i = p_i, \forall i \in K$. Let $V \subseteq Z$ be given by $V_j := V'_j$ for $j \notin K$ and by $V_i := Z_i$ for the other components. It is enough to show that V would also be eliminated via Algorithm 5 using $\tilde{f}$ (i.e. that V cannot contain any global optimizer).

First, assume that V' is eliminated by the method of active areas. Notice that from (8.3) each area reduction is made only between components $j_1, j_2 \notin K$. Clearly, if a component V'_s can be completely removed by applying such regions, then it could also be eliminated in V by the same test, since in V all the given reducing regions of V' are also present.

Secondly, consider the case when V' is deleted by the cutoff test. We first state two simple consequences of (8.3):

$$\mathrm{lb}(D(p_i,Z_j)) > \tilde{f} \Rightarrow \mathrm{ub}(D(p_i,V'_j)) = \mathrm{ub}(D(V'_i,V'_j)) > \tilde{f}, \text{ and} \tag{8.4}$$

$$\mathrm{lb}(D(p_i, Z_j)) > \tilde{f} \Rightarrow \mathrm{ub}(D(Z_i, V_j)) = \mathrm{ub}(D(V_i, V_j)) > \tilde{f} \qquad (8.5)$$

for all $i \in K,\ j \in \{1, \ldots, n\},\ i \neq j,\ V_j' \subseteq Z_j$, and $V_j \subseteq Z_j$.

Both statements can be easily proved in an indirect way by exploiting the fact that D is an inclusion function of the distance function d. Assuming the converse of the conclusions of (8.4) and (8.5), respectively, we could then say that $d(p_i, (x_j, y_j)) \leq \tilde{f}$, where (x_j, y_j) is arbitrarily chosen from V_j' for (8.4) and from V_j for (8.5). But due to $(x_j, y_j) \in Z_j$, this contradicts the premise of both (8.4) and (8.5).

Since V' is deleted by the cutoff test, we obtain

$$\tilde{f} > \mathrm{ub}(F(V')) = \min_{1 \leq i \neq j \leq n} \mathrm{ub}(D(V_i', V_j')) = \min_{\substack{1 \leq i \neq j \leq n \\ i \notin K, j \notin K}} \mathrm{ub}(D(V_i', V_j'))$$

$$= \min_{\substack{1 \leq i \neq j \leq n \\ i \notin K, j \notin K}} \mathrm{ub}(D(V_i, V_j)) = \min_{1 \leq i \neq j \leq n} \mathrm{ub}(D(V_i, V_j)) = \mathrm{ub}(F(V)).$$

Here, the conclusions of (8.4) and (8.5) are applied in the second and fourth equations, respectively, while the third equation follows from the construction of V. Thus, from the first and last statements of the chain we see that V can also be deleted using the same cutoff value. This completes the proof. □

From our computational experiences we found that the use of single values instead of regions was indispensable in achieving high precision verified solutions for the circle packing problems. The above method is performed in practice in the following way: for each component $(X_i, Y_i),\ i \in \{1, \ldots, n\}$ a stochastic search is applied to find a machine representable point $p_i \in (X_i, Y_i)$ for which $g(i) := \min_{1 \leq j \leq n, j \neq i} \mathrm{lb}(D(p_i, (X_j, Y_j)))$ is maximal. The search procedure we used was the stochastic optimization method GLOBAL [16]. If $g(i) > \mathrm{ub}(F_n(X, Y)) \geq \tilde{f}$ holds for the best $g(i)$ value found, then it is ensured that (8.3) also holds. (Notice that each p_i obtained is reliable, since $g(i)$ is computed by interval arithmetic.)

8.4 Optimal packing of $n = 28, 29$, and 30 points

8.4.1 Hardware and software environment

The optimization procedure was carried out on a PC with a Pentium IV 1.4 GHz processor and 1 GB RAM, running under the Linux operating system. As in Chapter 7, the global optimization algorithm framework is based on the C–XSC Numerical Toolbox libraries [43], while the low-level interval arithmetic routines were implemented by using PROFIL/BIAS [54]. In addition to this the multiple precision extension of PROFIL [55] was also incorporated in the algorithm to determine $\tilde{f}$ and to have correctly rounded decimal numbers in those I/O routines which are used to communicate with the user. (During the intermediate phases of our algorithm we used binary I/O routines.)

8.4.2 The global procedure

The currently known best packings for the investigated cases were published in [36] (for $n = 28$ points) and in [86] (for $n = 29$ and 30). The solution points and the corresponding function values can be found in [110], where the resulting components are rounded to 13 decimal digits. In order to obtain higher precision for the initial values of $\tilde{f}$, the *structures* of the best packings were investigated. These structures describe which points are located on the sides of the square and which pairs of points have minimal distances. The structures are formally given by systems of equations. (The simplified systems for $n = 28, 29$ can be found in Section 9.2.) Since the exact solutions of these systems are not known for $n = 28, 29$, we solved them numerically by Maple to a precision of 20 decimal places. The guaranteed enclosures of the objective function values on these more precise approximate results were then determined by a simple function evaluation with multiple precision intermediate interval data.

In contrast to this, the coordinates of the best known packing of 30 points have the exact form of $(qd + \text{par}(p+1)(1-4d), pa)$, where $d = (20-\sqrt{10})/75$, $a = 1/5$, par denotes the parity function and $p \in \{0,\ldots,5\}$, $q \in \{0,\ldots,4\}$. Moreover, the maximum value is exactly d. An interval enclosure of d was found via a simple multiple-precision expression evaluation.

From the above function enclosures we obtained the following $\tilde{f}_0$ lower bounds:

$$\tilde{f}_{0,28} = 0.2305354936426673,$$
$$\tilde{f}_{0,29} = 0.2268829007442089,$$
$$\tilde{f}_{0,30} = 0.2245029645310881.$$

These values allowed us to split the search space into 7×6 uniform cells as initial tiles for each of the three problems.

The original point arrangement setting (Problem 6.1) was modified in two steps: first, we considered squared distances instead of distances between points (see Section 6.1) and then we enlarged the search space from $[0,1] \times [0,1]$ to $[0,42] \times [0,42]$—as recommended by the first part of Remark 8.10. At the same time we evaluated the corresponding $\tilde{f}$ values:

$$\tilde{f}_{28} = 93.7506267944766227,$$
$$\tilde{f}_{29} = 90.8034005467881010,$$
$$\tilde{f}_{30} = 88.9083890308478496,$$

which were used during the whole computation.

Since the global procedure ran almost identically for the three problem instances, we introduce it only in the example for $n = 29$. (In [68] a similar report is given for $n = 28$.) In any case, the detailed computational results are given for all three problems in Tables 8.1, 8.2, and 8.3, respectively. The results are grouped according to the successive processing phases.

Based on our previous ideas, with the exception of the last phase we stopped the B&B algorithm after a preset number of iterations on each input combination to achieve a good level of efficiency for the whole procedure. This number was 3 in the first three phases and 10,000 in the fourth (refining) phase.

Table 8.1. Computational details for solving the packing problem with $n = 28$ points. The two columns on the left show the name and size of the input sets for the global optimization process. These sets are full tile combinations in Phase 1, the resulting sets of Algorithm 10 in Phase 2, the resulting sets of Algorithm 11 in Phase 3, the aggregated resulting sets of Phase 3 in Phase 4, and the only box to be refined in Phase 5, respectively. The middle group of columns lists the CPU time (CPUt, in seconds), the number of function evaluations (NFE), and the number of executions of the method of active areas (NMAA) necessary while processing the elements of the corresponding S sets. The column set on the right shows the name and size of the output sets of the given global optimization method.

S	$\|S\|$	CPUt	NFE	NMAA	$\bar{S}$	$\|\bar{S}\|$
Phase 1						
$S^{14}_{1..3}$	116 280	958	54 846	60 249	$\bar{S}^{14}_{1..3}$	17 799
$S^{15}_{1..3}$	54 264	139	4 401	16 658	$\bar{S}^{15}_{1..3}$	1 082
$S^{16}_{1..3}$	20 349	24	27	5 253	$\bar{S}^{16}_{1..3}$	0
$S^{13}_{1..3}$	203 490	2 526	232 268	174 995	$\bar{S}^{13}_{1..3}$	77 852
Phase 2						
${}^{13}_{1..3}S^{17}_{1..4}$	588 703	57 072	4 641 519	3 004 399	${}^{13}_{1..3}\bar{S}^{17}_{1..4}$	390 402
${}^{13}_{1..3}S^{18}_{1..4}$	632 289	47 620	2 843 863	2 214 934	${}^{13}_{1..3}\bar{S}^{18}_{1..4}$	226 564
${}^{13}_{1..3}S^{19}_{1..4}$	43 729	1 174	56 620	81 812	${}^{13}_{1..3}\bar{S}^{19}_{1..4}$	3 044
${}^{13}_{1..3}S^{20}_{1..4}$	0	-	-	-	${}^{13}_{1..3}\bar{S}^{20}_{1..4}$	0
${}^{14}_{1..3}S^{17}_{1..4}$	438 207	35 565	2 734 521	1 894 823	${}^{14}_{1..3}\bar{S}^{17}_{1..4}$	231 820
${}^{14}_{1..3}S^{18}_{1..4}$	354 167	22 466	1 312 522	1 090 923	${}^{14}_{1..3}\bar{S}^{18}_{1..4}$	103 582
${}^{14}_{1..3}S^{19}_{1..4}$	88 508	2 909	143 263	177 416	${}^{14}_{1..3}\bar{S}^{19}_{1..4}$	9 478
${}^{14}_{1..3}S^{20}_{1..4}$	2 540	13	12	2 550	${}^{14}_{1..3}\bar{S}^{20}_{1..4}$	0
${}^{14}_{1..3}S^{21}_{1..4}$	0	-	-	-	${}^{14}_{1..3}\bar{S}^{21}_{1..4}$	0
${}^{15}_{1..3}S^{17}_{1..4}$	21 844	1 276	95 333	74 306	${}^{15}_{1..3}\bar{S}^{17}_{1..4}$	7 816
${}^{15}_{1..3}S^{18}_{1..4}$	26 904	1 230	66 376	65 169	${}^{15}_{1..3}\bar{S}^{18}_{1..4}$	4 962
${}^{15}_{1..3}S^{19}_{1..4}$	14 134	312	13 370	22 543	${}^{15}_{1..3}\bar{S}^{19}_{1..4}$	853
${}^{15}_{1..3}S^{20}_{1..4}$	2 870	24	530	3 217	${}^{15}_{1..3}\bar{S}^{20}_{1..4}$	31
${}^{15}_{1..3}S^{21}_{1..4}$	0	-	-	-	${}^{15}_{1..3}\bar{S}^{21}_{1..4}$	0
${}^{15}_{1..3}S^{22}_{1..4}$	0	-	-	-	${}^{15}_{1..3}\bar{S}^{22}_{1..4}$	0
Phase 3						
${}^{13}_{1..3}S^{28}_{1..6}$	243 762	2 147	6 910	249 347	${}^{13}_{1..3}\bar{S}^{28}_{1..6}$	56
${}^{14}_{1..3}S^{28}_{1..6}$	998 204	9 107	38 614	1 028 665	${}^{14}_{1..3}\bar{S}^{28}_{1..6}$	506
Phase 4						
$S^{28}_{1..6}$	562	4 248	213 692	125 743	$\bar{S}^{28}_{1..6}$	6
Phase 5						
$S^{28}_{1..6}$	1	1 854	139 334	72 982	$\bar{S}^{28}_{1..6}$	1
$\sum$	3 850 807	190 664	12 598 021	10 365 984		1 075 854

Phase 1: First we evaluated $\bar{S}^{m}_{1..3}$ for some m. The initial sets $S^{m}_{1..3}$ were constructed from full tile combinations, as we had no previous information about the possible configurations. Notice that it is not worth evaluating the above sets when m is 'small', since we may not achieve significant reductions in the active areas for such m values. On the other hand, $\bar{S}^{m_0}_{1..3} = \emptyset$ implies $\bar{S}^{m}_{1..3} = \emptyset \; \forall m \geq m_0$ by

Table 8.2. Computational details for solving the packing problem with $n = 29$ points. The table is arranged in the same way as Table 8.1.

S	$\|S\|$	CPUt	NFE	NMAA	$\bar{S}$	$\|\bar{S}\|$
Phase 1						
$S^{15}_{1..3}$	54 264	305	14 928	22 758	$\bar{S}^{15}_{1..3}$	4 194
$S^{16}_{1..3}$	20 349	50	395	5 518	$\bar{S}^{16}_{1..3}$	40
$S^{17}_{1..3}$	5 985	8	0	1 574	$\bar{S}^{17}_{1..3}$	0
$S^{14}_{1..3}$	116 280	1 618	104 620	86 973	$\bar{S}^{14}_{1..3}$	33 794
$S^{13}_{1..3}$	203 490	3 454	338 771	230 121	$\bar{S}^{13}_{1..3}$	112 824
Phase 2						
${}^{13}_{1..3}S^{18}_{1..4}$	318 042	42 945	2 660 219	1 712 120	${}^{13}_{1..3}\bar{S}^{18}_{1..4}$	221 087
${}^{13}_{1..3}S^{19}_{1..4}$	111 682	8 323	475 178	391 266	${}^{13}_{1..3}\bar{S}^{19}_{1..4}$	35 833
${}^{13}_{1..3}S^{20}_{1..4}$	0	-	-	-	${}^{13}_{1..3}\bar{S}^{20}_{1..4}$	0
${}^{14}_{1..3}S^{18}_{1..4}$	691 320	77 284	4 638 570	3 182 171	${}^{14}_{1..3}\bar{S}^{18}_{1..4}$	383 148
${}^{14}_{1..3}S^{19}_{1..4}$	248 188	18 399	1 049 402	848 878	${}^{14}_{1..3}\bar{S}^{19}_{1..4}$	80 553
${}^{14}_{1..3}S^{20}_{1..4}$	20 685	492	15 298	31 861	${}^{14}_{1..3}\bar{S}^{20}_{1..4}$	520
${}^{14}_{1..3}S^{21}_{1..4}$	0	-	-	-	${}^{14}_{1..3}\bar{S}^{21}_{1..4}$	0
${}^{15}_{1..3}S^{18}_{1..4}$	113 761	10 142	605 305	448 297	${}^{15}_{1..3}\bar{S}^{18}_{1..4}$	49 428
${}^{15}_{1..3}S^{19}_{1..4}$	73 971	4 357	241 322	216 327	${}^{15}_{1..3}\bar{S}^{19}_{1..4}$	18 064
${}^{15}_{1..3}S^{20}_{1..4}$	20 303	605	21 069	34 290	${}^{15}_{1..3}\bar{S}^{20}_{1..4}$	1 137
${}^{15}_{1..3}S^{21}_{1..4}$	652	3	2	654	${}^{15}_{1..3}\bar{S}^{21}_{1..4}$	0
${}^{15}_{1..3}S^{22}_{1..4}$	0	-	-	-	${}^{15}_{1..3}\bar{S}^{22}_{1..4}$	0
${}^{16}_{1..3}S^{18}_{1..4}$	689	33	1 839	1 742	${}^{16}_{1..3}\bar{S}^{18}_{1..4}$	130
${}^{16}_{1..3}S^{19}_{1..4}$	855	20	999	1 471	${}^{16}_{1..3}\bar{S}^{19}_{1..4}$	60
${}^{16}_{1..3}S^{20}_{1..4}$	557	4	55	601	${}^{16}_{1..3}\bar{S}^{20}_{1..4}$	0
${}^{16}_{1..3}S^{21}_{1..4}$	114	0	0	114	${}^{16}_{1..3}\bar{S}^{21}_{1..4}$	0
${}^{16}_{1..3}S^{22}_{1..4}$	0	-	-	-	${}^{16}_{1..3}\bar{S}^{22}_{1..4}$	0
${}^{16}_{1..3}S^{23}_{1..4}$	0	-	-	-	${}^{16}_{1..3}\bar{S}^{23}_{1..4}$	0
Phase 3						
${}^{13}_{1..3}S^{29}_{1..6}$	2 349	17	67	2 400	${}^{13}_{1..3}\bar{S}^{29}_{1..6}$	1
${}^{14}_{1..3}S^{29}_{1..6}$	860 709	6 875	21 410	878 061	${}^{14}_{1..3}\bar{S}^{29}_{1..6}$	180
Phase 4						
$S^{29}_{1..6}$	181	3 785	157 168	80 731	$\bar{S}^{29}_{1..6}$	4
Phase 5						
$S^{29}_{1..6}$	1	3	166	107	$\bar{S}^{29}_{1..6}$	1
$\sum$	2 864 427	178 722	10 346 783	8 178 035		940 998

Corollary 8.9. We can utilize the above observations if we evaluate the sequence $\bar{S}^{\lceil n/2\rceil}_{1..3}, \bar{S}^{\lceil n/2\rceil+1}_{1..3}, \ldots, \bar{S}^{\lceil n/2\rceil+t}_{1..3}$ until we obtain $\bar{S}^{\lceil n/2\rceil+t}_{1..3} = \emptyset$. Then we evaluate the sets $\bar{S}^{\lfloor n/2\rfloor}_{1..3}$ (only if n is odd), $\bar{S}^{\lfloor n/2\rfloor-1}_{1..3}, \ldots, \bar{S}^{\lfloor n/2\rfloor-t+1}_{1..3}$. For $n = 29$, $\lceil n/2\rceil = 15$ and $\lfloor n/2\rfloor = 14$. Thus only $\bar{S}^{M_1}_{1..3}$, $M_1 = \{13, 14, 15, 16, 17\}$ were computed because $\bar{S}^{17}_{1..3}$ has proved to be empty. During the subsequent phases the sets $\bar{S}^{m}_{1..3}$, for $m \leq 12$ were regarded as consisting of full tiles.

Table 8.3. Computational details for solving the packing problem for $n = 30$ points. The table is arranged in the same way as Table 8.1.

S	$\lvert S\rvert$	CPUt	NFE	NMAA	$\bar{S}$	$\lvert\bar{S}\rvert$
			Phase 1			
$S^{15}_{1..3}$	54 264	499	27 566	30 132	$\bar{S}^{15}_{1..3}$	7 974
$S^{16}_{1..3}$	20 349	91	1 561	6 318	$\bar{S}^{16}_{1..3}$	255
$S^{17}_{1..3}$	5 985	13	20	1 590	$\bar{S}^{17}_{1..3}$	0
$S^{14}_{1..3}$	116 280	2 248	146 498	109 752	$\bar{S}^{14}_{1..3}$	47 003
			Phase 2			
${}^{14}_{1..3}S^{18}_{1..4}$	0	-	-	-	${}^{14}_{1..3}\bar{S}^{18}_{1..4}$	0
${}^{14}_{1..3}S^{19}_{1..4}$	414 511	40 089	2 298 205	1 685 461	${}^{14}_{1..3}\bar{S}^{19}_{1..4}$	182 441
${}^{14}_{1..3}S^{20}_{1..4}$	53 774	3 110	117 194	129 588	${}^{14}_{1..3}\bar{S}^{20}_{1..4}$	7 146
${}^{14}_{1..3}S^{21}_{1..4}$	0	-	-	-	${}^{14}_{1..3}\bar{S}^{21}_{1..4}$	0
${}^{15}_{1..3}S^{18}_{1..4}$	53 490	7 289	432 161	281 500	${}^{15}_{1..3}\bar{S}^{18}_{1..4}$	37 167
${}^{15}_{1..3}S^{19}_{1..4}$	191 328	14 536	814 390	653 867	${}^{15}_{1..3}\bar{S}^{19}_{1..4}$	63 213
${}^{15}_{1..3}S^{20}_{1..4}$	58 757	3 602	133 421	141 579	${}^{15}_{1..3}\bar{S}^{20}_{1..4}$	8 812
${}^{15}_{1..3}S^{21}_{1..4}$	3 899	45	564	4 357	${}^{15}_{1..3}\bar{S}^{21}_{1..4}$	7
${}^{15}_{1..3}S^{22}_{1..4}$	0	-	-	-	${}^{15}_{1..3}\bar{S}^{22}_{1..4}$	0
${}^{16}_{1..3}S^{18}_{1..4}$	3 055	311	17 255	12 708	${}^{16}_{1..3}\bar{S}^{18}_{1..4}$	1 359
${}^{16}_{1..3}S^{19}_{1..4}$	7 170	368	19 367	18 519	${}^{16}_{1..3}\bar{S}^{19}_{1..4}$	1 359
${}^{16}_{1..3}S^{20}_{1..4}$	4 697	153	4 878	7 983	${}^{16}_{1..3}\bar{S}^{20}_{1..4}$	263
${}^{16}_{1..3}S^{21}_{1..4}$	1 186	13	206	1 342	${}^{16}_{1..3}\bar{S}^{21}_{1..4}$	7
${}^{16}_{1..3}S^{22}_{1..4}$	20	0	0	20	${}^{16}_{1..3}\bar{S}^{22}_{1..4}$	0
${}^{16}_{1..3}S^{23}_{1..4}$	0	-	-	-	${}^{16}_{1..3}\bar{S}^{23}_{1..4}$	0
			Phase 3			
${}^{14}_{1..3}S^{30}_{1..6}$	16 799	104	10	16 807	${}^{14}_{1..3}\bar{S}^{30}_{1..6}$	0
${}^{15}_{1..3}S^{30}_{1..6}$	239 430	2 068	3 197	242 002	${}^{15}_{1..3}\bar{S}^{30}_{1..6}$	36
			Phase 4			
$S^{30}_{1..6}$	36	1 391	79 916	39 935	$\bar{S}^{30}_{1..6}$	2
			Phase 5			
$S^{30}_{1..6}$	1	0	18	7	$\bar{S}^{30}_{1..6}$	1
$\sum$	1 245 031	75 930	4 096 427	3 383 467		357 045

Acceleration of Phase 1. We reduced the necessary computational effort by filtering out symmetric cases from each set $S^{m}_{1..3}$. Next we applied the following transformations: reflections to the axis $x = 10.5$ and $y = 21$ and a reflection about the point $(10.5, 21)$. Each full tile combination was represented by a binary string. Hence the filtering procedure could be performed applying simple transformations on these strings. After the elimination procedure was done on $S^{m}_{1..3}$, the inverse transformations had to be applied for the remaining areas to generate the correct $\bar{S}^{m}_{1..3}$.

Phase 2: At this stage we had $\bar{S}^{m}_{1..3}$ for $0 \le m \le 21$. We evaluated ${}^{m}_{1..3}\bar{S}^{m+i}_{1..4}$ for each necessary (m, i) pairs in two steps. First, we computed ${}^{m}_{1..3}S^{m+i}_{1..4}$ by Algorithm 10 and then ran the optimization algorithm on these sets to obtain ${}^{m}_{1..3}\bar{S}^{m+i}_{1..4}$. The parameters m and i were determined in the following way (cf. Remark 8.12):

We had to deal only with $m \in M_2 = \{13, 14, 15, 16\}$, where M_2 was the result of Phase 1; i.e. we added one column to each element of the set $\bar{S}^{M_2}_{1..3}$. According to Remark 8.12, a lower bound m_l for the number of output tiles $m + i$ could be determined by a process similar to Phase 1. $\bar{S}^{12}_{5..6}$ proved to be empty (but $\bar{S}^{11}_{5..6}$ did not), so we had to choose at least $m_l = 18$ tiles from the first four columns. The upper bounds for $m + i$ were set to $m + 7$ for each m.

Phase 3: It is easy to see that we could proceed by adding other columns to the resulting patterns of Phase 2. Instead, we found that we already had enough local information at the end of this phase to join our remaining combinations by Algorithm 11 and to obtain a relatively small set of combinations consisting of 29 tiles.

As a result of Phase 2, we knew ${}^{\,i}_{1..3}\bar{S}^{m_i}_{1..4}$ for all $18 \le m_i \le 29$, $i \in M_2$. (Recall that we did not need to consider m_i values with $i \le m_i < 18$.) Since the above sets are non-empty only when $m_{13} \in M_{13} = \{18, 19\}$, $m_{14} \in M_{14} = \{18, 19, 20\}$, $m_{15} \in M_{15} = \{18, 19, 20\}$ and $m_{16} \in M_{16} = \{18, 19\}$, respectively, it was enough to consider these m_i values when applying Algorithm 11. As was suggested in the discussion of Algorithm 11, the other input sets ${}^{\,i}_{4..6}\bar{S}^{M_i}_{3..6}$, $i \in M_2$ could be evaluated from ${}^{\,i}_{1..3}\bar{S}^{M_i}_{1..4}$ by rotating each element 180 degrees about the midpoint $(21, 21)$ of the square.

Hence in Phase 3 we had to join ${}^{13}_{1..3}\bar{S}^{M_{13}}_{1..4}$ with ${}^{16}_{4..6}\bar{S}^{M_{16}}_{3..6}$ (this process is called $Join(13, 16)$ in the sequel) and, similarly, join ${}^{14}_{1..3}\bar{S}^{M_{14}}_{1..4}$ with ${}^{15}_{4..6}\bar{S}^{M_{15}}_{3..6}$ ($Join(14, 15)$), then join ${}^{15}_{1..3}\bar{S}^{M_{15}}_{1..4}$ with ${}^{14}_{4..6}\bar{S}^{M_{14}}_{3..6}$ ($Join(15, 14)$) and, finally, join ${}^{16}_{1..3}\bar{S}^{M_{16}}_{1..4}$ with ${}^{13}_{4..6}\bar{S}^{M_{13}}_{3..6}$ ($Join(16, 13)$). The resulting sets ${}^{M_2}_{1..3}S^{29}_{1..6}$ had to be processed by the optimization algorithm in sequential order. As we stated that, for a solution having an objective function value greater than or equal to $\tilde{f}$, only 13, 14, 15, or 16 tiles can be chosen from each half of the search region, the union of the resulting sets,

$$ {}^{13}_{1..3}\bar{S}^{29}_{1..6} \cup {}^{14}_{1..3}\bar{S}^{29}_{1..6} \cup {}^{15}_{1..3}\bar{S}^{29}_{1..6} \cup {}^{16}_{1..3}\bar{S}^{29}_{1..6} $$

can be regarded as $S^{29}_{1..6}$, i.e. a set which contains all the global maximizers.

The acceleration of Phase 3: Again, symmetry properties were applied to reduce the necessary computations. Here without loss of generality we may assume that in each element of $S^{29}_{1..6}$ the left half of the search square contains more active regions than the right one (keeping in mind that points located on the vertical bisecting line segment can be viewed as lying in both half squares). Consequently, we need not execute $Join(15, 14)$ and $Join(16, 13)$ when $Join(14, 15)$ and $Join(13, 16)$ have already been done. Another idea that exploits the symmetry properties is applied for even n numbers. For example, in $Join(14, 14)$ for $n = 28$ we assumed that column 3 does not contain fewer regions than column 4, i.e. $Join(14, 14)$ could be restricted to join each ${}^{14}_{1..3}\bar{S}^{i}_{1..4}$ with each ${}^{14}_{4..6}\bar{S}^{j}_{3..6}$ only when $j \ge i$. A similar acceleration was done for $n = 30$ in $Join(15, 15)$. (We presented only two very simple cases for filtering symmetric combinations, which can be easily checked in our implementation. Obviously, one can develop more sophisticated methods that result in larger filtering improvements.)

Table 8.4. The interval bounds of the global maximizer point for packing 28 points, the X_i coordinates.

i	X_i
1.	[0.0000000000000000, 0.0000000000000142]
2.	[0.0000000000000000, 0.0000000000000723]
3.	[0.0000000000000000, 0.0000000000000617]
4.	[0.0000000000000000, 0.0000000000000689]
5.	[0.0000000000000000, 0.0000000000000094]
6.	[0.1818703764471228, 0.1818703764471709]
7.	[0.1996495939685054, 0.1996495939687475]
8.	[0.1996495939685047, 0.1996495939686531]
9.	[0.1918713858351473, 0.1918713858351900]
10.	[0.3637407528942488, 0.3637407528964056]
11.	[0.3815199704156590, 0.3815199704157629]
12.	[0.3992990052060252, 0.4023815131388759]
13.	[0.3915209798036835, 0.3915209798037213]
14.	[0.3837427716702937, 0.3837427716703746]
15.	[0.5633903468627852, 0.5633903468641770]
16.	[0.5839312817244451, 0.5839312817244956]
17.	[0.5911705737722300, 0.5911705737722556]
18.	[0.5833923656388268, 0.5833923656389046]
19.	[0.7630399408313210, 0.7630399408318696]
20.	[0.7694645063572478, 0.7694645063573329]
21.	[0.7727203431310999, 0.7727203431311068]
22.	[0.7830419596073901, 0.7830419596074357]
23.	[0.7830419596073660, 0.7830419596074357]
24.	[0.9935754344739882, 0.9935754344745461]
25.	[0.9999999999999152, 1.0000000000000000]
26.	[0.9999999999999947, 1.0000000000000000]
27.	[0.9999999999999551, 1.0000000000000000]
28.	[0.9999999999999316, 1.0000000000000000]

Phase 4: In accordance with our previous remarks, only the two resulting sets ${}^{13}_{1..3}\bar{S}^{29}_{1..6}$ and ${}^{14}_{1..3}\bar{S}^{29}_{1..6}$ were put into a set $S^{29}_{1..6}$ to be refined. Since this set may contain a lot of equivalent solutions, we still did not use the original stopping criterion of Algorithm 5. Instead, we increased the maximal number of allowed iteration steps from 3 to 10000. As a result, we managed to reduce the number of remaining combinations to 6 for $n = 28$, to 4 for $n = 29$, and to 2 for $n = 30$.

Phase 5: After Phase 4 it was possible to check all the remaining regions one by one. (However, for some other n values a much bigger number of combinations may remain; thus, we are planning to automate this step in the future.) For the problem of packing 29 points the first two combinations yielded to symmetric packings (since their initial tile combinations were symmetric), as did the other two combinations. But we found that the initial tile sets of the first group could not be transformed to the tile sets of the second group of combinations. The reason for this is that the 7×6

Table 8.5. The interval bounds of the global maximizer point for packing 28 points, the Y_i coordinates.

i	Y_i
1.	[0.0000000000000000, 0.0000000000000182]
2.	[0.2833357019511142, 0.2833357019512377]
3.	[0.5138711955937813, 0.5138711955939050]
4.	[0.7444066892364485, 0.7444066892365722]
5.	[0.9999999999999859, 1.0000000000000000]
6.	[0.1416678509755570, 0.1416678509756246]
7.	[0.3986034487723758, 0.3986034487725656]
8.	[0.6291389424150430, 0.6291389424152327]
9.	[0.8722033446182193, 0.8722033446182866]
10.	[0.0000000000000000, 0.0000000000015377]
11.	[0.2569355977967746, 0.2569355977969572]
12.	[0.5089492733445356, 0.5138712298523256]
13.	[0.7569355977968880, 0.7569355977969508]
14.	[0.9999999999999370, 1.0000000000000000]
15.	[0.1152677468208971, 0.1152677468228100]
16.	[0.3672817571470099, 0.3672817571470896]
17.	[0.6416678509755755, 0.6416678509756131]
18.	[0.8847322531786048, 0.8847322531787316]
19.	[0.0000000000000000, 0.0000000000001518]
20.	[0.2304459563257084, 0.2304459563258553]
21.	[0.4995893688792053, 0.4995893688792224]
22.	[0.7694645063572883, 0.7694645063573327]
23.	[0.9999999999999551, 1.0000000000000000]
24.	[0.0000000000000000, 0.0000000000000352]
25.	[0.2304459563257084, 0.2304459563257429]
26.	[0.4609814499683757, 0.4609814499684068]
27.	[0.6915169436110431, 0.6915169436111594]
28.	[0.9220524372537104, 0.9220524372539002]

splitting is not invariant under rotations by ± 90 degrees around the midpoint of the square.

However, it was not hard to see that if we split the square into 6×7 regular tiles, each remaining region of one of the latter two combinations was located in different tiles of the new splitting, and the new tile set corresponding to this combination was symmetric with the patterns of the first two combinations. Consequently, it was enough to consider only one of the four combinations in further investigations. (One might suppose that four combinations should exist for both groups of remaining subproblems. Indeed, it is easy to check that the additional combinations were filtered during Phase 3 when the above-mentioned symmetry rules were applied.)

In the 28 and 30 point cases, similar observations were made, which had the same consequences, i.e. that one had to deal with only one combination. Here we obtained the first main result of our present study: *Let* $n \in \{28, 29, 30\}$. *Apart from symmetric*

Table 8.6. The interval bounds of the global maximizer point for packing 29 points, the X_i coordinates.

i	X_i
1.	$[0.0000000000000000, 0.0000000000000246]$
2.	$[0.0000000000000000, 0.0000000000000217]$
3.	$[0.0000000000000000, 0.0000000000000199]$
4.	$[0.0000000000000000, 0.0000000000000136]$
5.	$[0.0000000000000000, 0.0532481705058822]$
6.	$[0.2009548756284472, 0.2009548756284715]$
7.	$[0.2009548756284483, 0.2009548756284689]$
8.	$[0.2009548756284484, 0.2009548756284683]$
9.	$[0.2009548756284486, 0.2009548756284655]$
10.	$[0.2762399917698153, 0.2762399917698536]$
11.	$[0.4019097512568943, 0.4019097512569189]$
12.	$[0.4019097512568967, 0.4019097512569177]$
13.	$[0.4019097512568968, 0.4019097512569169]$
14.	$[0.4019097512568973, 0.4019097512569210]$
15.	$[0.5983961069856828, 0.5983961069857054]$
16.	$[0.5983961069856840, 0.5983961069857049]$
17.	$[0.5983961069856844, 0.5983961069857049]$
18.	$[0.5983961069856861, 0.5983961069857088]$
19.	$[0.5031228925140242, 0.5031228925140623]$
20.	$[0.7948824627144688, 0.7948824627144926]$
21.	$[0.7948824627144706, 0.7948824627144921]$
22.	$[0.7948824627144710, 0.7948824627144921]$
23.	$[0.7948824627144749, 0.7948824627144919]$
24.	$[0.7290826584835373, 0.7290826584837260]$
25.	$[0.9999999999999812, 1.0000000000000000]$
26.	$[0.9999999999999812, 1.0000000000000000]$
27.	$[0.9999999999999820, 1.0000000000000000]$
28.	$[0.9999999999999859, 1.0000000000000000]$
29.	$[0.9550424244530482, 0.9550424244532232]$

cases, one initial tile combination contains all the globally optimal solutions of the packing problem of n points. We can demonstrate the truth of this statement by proving that all the global solutions are located in a very small region within a given tile combination.

As a further refinement of the only remaining box, the method for guessing free points and shrinking several components of the remaining regions was applied (see Section 8.3). We found that for $n = 28$ only the 12th component could be replaced by a point; for $n = 29$ only the 5th component; while for $n = 30$ no components could be replaced. (These facts seem to fit in with what we know about the location of the free points of the best-known packings.) The only remaining regions were then refined via Algorithm 5 using the stopping criterion parameter of $\varepsilon = 10^{-10}$. As expected, the cost of this local investigation was relatively small compared to the global part of

Table 8.7. The interval bounds of the global maximizer point for packing 29 points, the Y_i coordinates.

i	Y_i
1.	$[0.0000000000000000, 0.0000000000000393]$
2.	$[0.2268829007442089, 0.2268829007442435]$
3.	$[0.4537658014884179, 0.4537658014884444]$
4.	$[0.6806487022326268, 0.6806487022326528]$
5.	$[0.9073423366158249, 1.0000000000000000]$
6.	$[0.1053232576939058, 0.1053232576939296]$
7.	$[0.3322061584381170, 0.3322061584381365]$
8.	$[0.5590890591823258, 0.5590890591823453]$
9.	$[0.7859719599265345, 0.7859719599265537]$
10.	$[0.9999999999999868, 1.0000000000000000]$
11.	$[0.0000000000000000, 0.0000000000000215]$
12.	$[0.2268829007442089, 0.2268829007442285]$
13.	$[0.4537658014884179, 0.4537658014884354]$
14.	$[0.6806487022326268, 0.6806487022326512]$
15.	$[0.1134414503720853, 0.1134414503721164]$
16.	$[0.3403243511162964, 0.3403243511163233]$
17.	$[0.5672072518605076, 0.5672072518605315]$
18.	$[0.7940901526047168, 0.7940901526047401]$
19.	$[0.9999999999999830, 1.0000000000000000]$
20.	$[0.0000000000000000, 0.0000000000000197]$
21.	$[0.2268829007442089, 0.2268829007442256]$
22.	$[0.4537658014884178, 0.4537658014884334]$
23.	$[0.6806487022326267, 0.6806487022326419]$
24.	$[0.9795541003348881, 0.9795541003356186]$
25.	$[0.0969672447170968, 0.0969672447171463]$
26.	$[0.3238501454613097, 0.3238501454613550]$
27.	$[0.5507330462055206, 0.5507330462055635]$
28.	$[0.7776159469497363, 0.7776159469497720]$
29.	$[0.9999999999999683, 1.0000000000000000]$

our whole procedure. In the end after transforming the enclosures of the result boxes back to the original point packing problem, we obtained the tightest currently known enclosures of the global maximum values. These are

$$\begin{aligned} F^*_{28} &= [0.2305354936426673, 0.2305354936426743],\ w(F^*_{28}) \approx 7 \cdot 10^{-15}, \\ F^*_{29} &= [0.2268829007442089, 0.2268829007442240],\ w(F^*_{29}) \approx 2 \cdot 10^{-14}, \\ F^*_{30} &= [0.2245029645310881, 0.2245029645310903],\ w(F^*_{30}) \approx 2 \cdot 10^{-15}. \end{aligned}$$

The enclosures of the particular global maximizers $(X, Y)^*_n$ are listed in Tables 8.4, 8.5, 8.6, 8.7, 8.8, and 8.9 respectively. The precision of the components are highlighted by underscores. Our conclusions are the following:

Apart from symmetric cases, all the global optimizer points of the packing problem of n points (where $n \in \{28, 29, 30\}$) are located in the reported boxes. These boxes have very tight components with the obvious exception of the components en-

Table 8.8. The interval bounds of the global maximizer point for packing 30 points, the X_i coordinates.

i	X_i
1.	[0.1019881418756426, 0.1019881418756476]
2.	[0.0000000000000000, 0.0000000000000026]
3.	[0.1019881418756451, 0.1019881418756477]
4.	[0.0000000000000000, 0.0000000000000025]
5.	[0.1019881418756451, 0.1019881418756478]
6.	[0.0000000000000000, 0.0000000000000045]
7.	[0.3264911064067307, 0.3264911064067357]
8.	[0.2245029645310881, 0.2245029645310907]
9.	[0.3264911064067332, 0.3264911064067358]
10.	[0.2245029645310881, 0.2245029645310905]
11.	[0.3264911064067332, 0.3264911064067359]
12.	[0.2245029645310881, 0.2245029645310926]
13.	[0.4490059290621762, 0.4490059290621788]
14.	[0.4490059290621762, 0.4490059290621786]
15.	[0.4490059290621762, 0.4490059290621807]
16.	[0.5509940709378189, 0.5509940709378239]
17.	[0.5509940709378213, 0.5509940709378240]
18.	[0.5509940709378213, 0.5509940709378240]
19.	[0.7754970354689073, 0.7754970354689120]
20.	[0.6735088935932643, 0.6735088935932668]
21.	[0.7754970354689095, 0.7754970354689120]
22.	[0.6735088935932643, 0.6735088935932666]
23.	[0.7754970354689095, 0.7754970354689120]
24.	[0.6735088935932643, 0.6735088935932688]
25.	[0.9999999999999955, 1.0000000000000000]
26.	[0.8980118581243525, 0.8980118581243549]
27.	[0.9999999999999976, 1.0000000000000000]
28.	[0.8980118581243525, 0.8980118581243548]
29.	[0.9999999999999977, 1.0000000000000000]
30.	[0.8980118581243525, 0.8980118581243570]

closing the possible free points for $n = 28, 29$. Furthermore, the reported enclosures confirm that the structure of the currently known best packings are optimal within the precision of $w(F_n^*)$. In other words, all the exact global optimizers are located in $(X,Y)_{28}^*$, $(X,Y)_{29}^*$, and $(X,Y)_{30}^*$, and the exact global optima differ from the currently best-known function values by at most $w(F_n^*)$.

It is also worth mentioning that our verified procedure decreased the uncertainty in the location of the optimal packing by as much as

$$\log\left(w([0,1])^{2n} / \prod_{i=1}^{n} w(X_n^*)w(Y_n^*)\right),$$

i.e. by more than 711, 764, and 872 orders of magnitude, respectively.

Table 8.9. The interval bounds of the global maximizer point for packing 30 points, the Y_i coordinates.

i	Y_i
1.	[0.0000000000000000, 0.0000000000000026]
2.	[0.1999999999999995, 0.2000000000000022]
3.	[0.3999999999999989, 0.4000000000000017]
4.	[0.5999999999999984, 0.6000000000000011]
5.	[0.7999999999999978, 0.8000000000000006]
6.	[0.9999999999999972, 1.0000000000000000]
7.	[0.0000000000000000, 0.0000000000000028]
8.	[0.1999999999999995, 0.2000000000000022]
9.	[0.3999999999999989, 0.4000000000000017]
10.	[0.5999999999999984, 0.6000000000000011]
11.	[0.7999999999999978, 0.8000000000000006]
12.	[0.9999999999999972, 1.0000000000000000]
13.	[0.1999999999999995, 0.2000000000000021]
14.	[0.5999999999999984, 0.6000000000000011]
15.	[0.9999999999999972, 1.0000000000000000]
16.	[0.0000000000000000, 0.0000000000000025]
17.	[0.3999999999999989, 0.4000000000000016]
18.	[0.7999999999999978, 0.8000000000000006]
19.	[0.0000000000000000, 0.0000000000000024]
20.	[0.1999999999999995, 0.2000000000000020]
21.	[0.3999999999999990, 0.4000000000000015]
22.	[0.5999999999999986, 0.6000000000000011]
23.	[0.7999999999999981, 0.8000000000000006]
24.	[0.9999999999999976, 1.0000000000000000]
25.	[0.0000000000000000, 0.0000000000000023]
26.	[0.1999999999999995, 0.2000000000000019]
27.	[0.3999999999999991, 0.4000000000000015]
28.	[0.5999999999999987, 0.6000000000000011]
29.	[0.7999999999999981, 0.8000000000000006]
30.	[0.9999999999999976, 1.0000000000000000]

As Tables 8.1 to 8.3 show, the total computational time of the optimality proof was approximately 53, 50, and 21 hours for the packing of 28, 29, and 30 points, respectively. Note that the CPU time of those parts of the method not given in the tables (such as the determination of m_l in Phase 2 and the execution time of Algorithms 10 and 11) took about only one extra minute for each problem instance. The total time complexities are considerably less than the forecasted execution times (i.e. one or even more decades) suggested by the earlier methods. The memory complexity of our procedure was small in most parts of the method, since the results of the intermediate phases could be saved to storage devices. Moreover, due to the sequential and rapid processing of the tile combinations, the main optimization method required only a small amount of memory. In several cases, the memory requirement for Algorithms

10 and 11 was large for large sets of input combinations, but our main memory was still big enough to store all the necessary combinations at a time to avoid swapping.

8.5 Summary of the results for the advanced algorithm

In this chapter we introduced an advanced variant of our verified, interval arithmetic-based optimization method meant for solving circle (point) packing problems. Each part of our method was discussed via detailed algorithmic descriptions and proofs of correctness. As our final results show, we solved the previously unsolved problem instances of packing 28, 29, and 30 points. We also demonstrated that the previously best found configurations are globally optimal within very tight tolerance values. We listed the box values enclosing the exact optimizers, and concluded that the current best function value differs from the maximum by at most $7 \cdot 10^{-15}$, $2 \cdot 10^{-14}$, and $2 \cdot 10^{-15}$ for $n = 28, 29$, and 30, respectively. This seems to be a remarkable achievement.

When compared to the first interval method of Chapter 7, the reader saw that our accelerated method contains two key procedures. First, we developed a more efficient area reduction algorithm using a polygon representation. Second, a very effective technique based on the subsets of tile combinations was introduced for handling and eliminating groups of subproblems together. As a result, the total computational time was dramatically smaller than the predicted time for running the earlier methods.

The basic elements of the proposed method will most probably play a role in further studies of both circle and point arrangement problems for $n \geq 31$ and perhaps also for the generalizations of the problem class. For the next unsolved problem case, $n = 31$, a 8×6 tile splitting can be applied, which will probably require an additional intermediate phase for 'growing' the dimensionality of the subproblems in question. Thus we are planning to further develop our accelerated method to eliminate tile combinations in a more general way. The first part of this work would consist of programming features like extending Algorithm 10 so that it has the ability to add new columns of tiles, and modifying Algorithm 11 to handle general intersecting tile positions of the combinations to be joined. By employing this more general approach, more sophisticated multi-phase methods could be designed, reducing the total cost of processing intermediate tile combinations, and opening the way for solving much harder point and circle packing problems and their generalizations.

9

Interval Methods for Verifying Structural Optimality

In the previous chapters we created a problem-specific optimization method based exclusively on interval arithmetic calculations and solved the previously open problems of packing 28, 29, and 30 points in the *numerical sense*. In other words, we gave tight bounds for all the optimal solutions and for the optimum values. In this chapter we complete the computer-assisted optimality proof of these cases: we determine all the optimal solutions in the *geometric sense*. Namely, we prove that the currently best-known packing *structures* result in optimal packings and moreover apart from symmetric configurations and the movement of well-identified free circles, these are the only optimal packings [69]. The required statements will be verified with mathematical rigor using interval arithmetic tools.

9.1 Packing structures and previous numerical results

The structure of a point arrangement configuration (or an equivalent circle packing one) describes

(a) which points are located on the sides of the square (which circles are touching one side of the square),

(b) which pairs of points have the minimal distance (which circles are touching each other), and

(c) which are the free points[1] (free circles) of the packing.

Obviously, the question of whether the distance between two 'machine points' (given by two binary floating-point numbers) is exactly the minimal pairwise distance cannot be answered by simply calculating distances or interval enclosures of distances on a computer. This fact produces the main obstacle in proving structural properties.

[1] A free point of a packing configuration can be slightly moved without affecting the minimal pairwise distance, i.e. the objective function value. See Definition 7.4 in Section 7.2 for the precise meaning used throughout this work.

Using the notation introduced in Section 6.2, from the numerical results of Section 8.4 we have the following numerical information in hand for $n = 28, 29$, and 30.

- An interval enclosure of all global optimizers (apart from symmetric cases) can be denoted by $(X, Y)_n^* = (X_1, \ldots, X_n, Y_1, \ldots, Y_n) \subset [0, 1]^{2n}$. Each component of $(X, Y)_n^*$ has a width of $\approx 10^{-13} - 10^{-15}$, with the exception of components enclosing a possibly free point for $n = 28, 29$.
- An enclosure of the global optimum value is $F_n^* \in \mathbb{I}$. The width of this interval is about 10^{-14}.
- For $n = 28$ and 29, an enclosure $(X_{free}, Y_{free})_n \subset (X_k, Y_k)_n^*$ of a point (x_{free}, y_{free}), such that
$$\mathrm{lb}(D((X_{free}, Y_{free})_n, (X_j', Y_j')_n)) > \mathrm{ub}(F^*) \geq f^* \tag{9.1}$$
for $(X_j, Y_j)_n^* \subseteq (X_j', Y_j')_n$, and for all $j = 1, \ldots, n$, $j \neq k$. $(X_{free}, Y_{free})_n$ is used to temporarily reduce those components of the search boxes which probably enclose free points, and thus prevent the B&B algorithm from performing unnecessary subdivisions on these components.
- A real-type, high-precision approximate solution computed from a configuration that is thought to be an optimal packing is $(x_{appr}, y_{appr})_n \in [0, 1]^{2n}$.

To simplify the notation, from now on we shall neglect the lower n indices unless they are explicitly required.

The aforementioned best available packing structures were discovered in [36] for $n = 28$ and in [86] for $n = 29$ and 30, respectively. The graphical representations of these structures are—as we shall soon prove—identical to those shown in Figure 9.2.

We use the term *rigid point* to refer to that point of a packing which is not free (or in the case of interval enclosures, to that which which we conjecture to be fixed). The component vectors corresponding to the (conjectured) rigid points will be denoted by an upper r index.

The present verification procedures were implemented by again using the PROFIL/BIAS v.2.0 interval package [56]. The source codes and the numerical results are available at `http://www.sztaki.hu/~markot/packcirc.htm`.

Since the numerical results of Section 8.4 were also used in the present verification procedures, special care had to be taken to communicate these data items in an error-free way between programs. To achieve this goal, we applied the *binary representation of interval vectors* provided by the complete algorithm of Section 8.4 to pass data between successive stages. In addition to this, we employed the `Enclosure()` function of the interval package [56], which returns the interval enclosure of an interval given by a *string*.

In the following section three assertions and the corresponding computer-assisted verifications will be given to prove the optimality and uniqueness of the supposedly best packing structures.

9.2 Optimality and uniqueness properties

Assertion 1 *For* $n = 28, 29,$ *and* 30, *the system of equations describing the rigid part of the best-known packing structure has exactly one* $(\bar{x}, \bar{y})^r$ *solution in the particular components* $(X, Y)^{*,r}$ *of* $(X, Y)^*$.

To verify this assertion, first the rigid structures of the best-known packings were determined in terms of the properties (a) and (b) of Section 9.1. These properties can be expressed as a quadratic system of equations using the coordinates of the points and the minimal pairwise distance.

For $n = 28$ and 29, the exact solutions of these systems are not known, so we had to solve them in a numerical way, once again using interval computations. We first performed some variable substitutions on the systems to achieve a reduction in the number of variables. Denoting the location of the ith point by $(x_i, y_i) \in [0, 1]^2$, and the minimal pairwise distance by p, for $n = 28$ the simplified (but still quadratic) system of equations was

$$
\begin{aligned}
p^2 &= (x_{10}/2)^2 + (y_2/2)^2 \\
p^2 &= (x_{23}/2 - \sqrt{3}p/2)^2 + (y_2/2 + p - 1/2)^2 \\
p^2 &= (x_{10}/2 - x_{19}/2)^2 + y_{15}^2 \\
p^2 &= (\sqrt{3}p/2 - x_{19}/2)^2 + (y_2/2 + p/2 - y_{15})^2 \\
p^2 &= (\sqrt{3}p/2 - x_{23}/2)^2 + (y_2/2 + p/2 - y_{18} + 1/2)^2 \\
p^2 &= (x_{19}/2 - x_{16})^2 + (y_{15} + y_2/2 - y_{16})^2 \\
p^2 &= (x_{16} - 1 + p)^2 + (y_{16} - y_{25})^2 \\
p^2 &= (x_{16} - x_{21})^2 + (y_{16} - y_{21})^2 \\
p^2 &= (x_{23}/2 + \sqrt{3}p/2 - x_{21})^2 + (1/2 + y_2/2 - y_{21})^2 \\
p^2 &= (x_{19} + p - 1)^2 + y_{25}^2 \\
p^2 &= (x_{21} - 1)^2 + (y_{21} - y_{25} - p)^2 \\
p^2 &= (x_{23} - 1)^2 + (1 - 3p - y_{25})^2,
\end{aligned}
$$

while for $n = 29$ it was much simpler:

$$
\begin{aligned}
p^2 &= x_6^2 + y_6^2 \\
p^2 &= (x_6 - x_{19} + p)^2 + (y_6 + 3p - 1)^2 \\
p^2 &= (2x_6 + \sqrt{3}p/2 - x_{19})^2 + (7p/2 - 1)^2 \\
p^2 &= (2x_6 + \sqrt{3}p/2 - x_{19}/2 - x_{29}/2)^2 + (7p/2 - y_{24})^2 \\
p^2 &= (x_{19}/2 - x_{29}/2)^2 + (1 - y_{24})^2 \\
p^2 &= (2x_6 + \sqrt{3}p - 1)^2 + y_{25}^2 \\
p^2 &= (1 - x_{29})^2 + (y_{25} + 3p - 1)^2.
\end{aligned}
$$

During the coding of the systems of equations, we also took special care of the reliable type of representation used for constants by applying proper interval enclosures.

The systems were then solved numerically via the interval Newton–Gauss–Seidel iteration method available in the interval Numerical Toolbox for C–XSC [43]. Besides furnishing reliable interval enclosures for each solution, this procedure is also able to verify the *existence* and *local / global uniqueness* of a solution. To demonstrate that these properties can be realized even for a large box, the search region was found by blowing up $(X,Y)^{*,r}$ and $\sqrt{F_n^*}$ to a width of 0.01. In spite of this, the interval Newton method was able to prove the existence and global uniqueness of a solution and to give its properly tight $(\bar{X},\bar{Y})^r$ enclosure. Finally, the verification of the inclusion property $(\bar{X},\bar{Y})^r \subseteq (X,Y)^{*,r}$ implying Assertion 1 was also successful.

Unlike the cases $n = 28, 29$, for $n = 30$ the solution of the system of equations is given in an exact analytic form (see Section 8.4 for more details).

Consequently, Assertion 1 could be proved in a straightforward manner by evaluating the guaranteed enclosures $(\bar{X},\bar{Y})^r$ of the exact coordinate values and then verifying $(\bar{X},\bar{Y})^r \subseteq (X,Y)^{*,r}$ as before.

The importance of Assertion 1 is highlighted by the fact that it allows one to associate a unique point arrangement (i.e. a solution vector) located in the box $(X,Y)^{*,r}$ with the rigid part of the guessed optimal structure (with a geometric solution).

Assertion 2 *For* $n = 28, 29$, *the solution* $(\bar{x},\bar{y})^r$ *of Assertion 1 can be extended by a free point located in that component* $(X_k,Y_k)^*$ *of* $(X,Y)^*$ *which is not considered in* $(X,Y)^{*,r}$.

Formally, Assertion 2 means that one has to prove the existence of a point $(x_k, y_k) \in (X_k, Y_k)^*$, for which

$$d((x_k,y_k),(\bar{x}_j,\bar{y}_j)) > f^* \tag{9.2}$$

holds for each $j = 1,\ldots,n,\ j \neq k$. Although we do not know the exact values of $(\bar{x}_j,\bar{y}_j)$ and f^*, we can test the existence of a small rectangle (X_k,Y_k) enclosing (x_k,y_k), such that $(X_k,Y_k) \subseteq (X_k,Y_k)^*$ and

$$\mathrm{lb}(D((X_k,Y_k),(\bar{X}_j,\bar{Y}_j))) > \mathrm{ub}(F^*) \tag{9.3}$$

for each $j = 1,\ldots,n,\ j \neq k$. Evidently, (9.3) implies (9.2). Comparing the latter equation with (9.1), (X_{free},Y_{free}) turns out to be a suitable value for (X_k,Y_k). In theory, since $(\bar{X}_j,\bar{Y}_j) \subseteq (X_j,Y_j)^* \subseteq (X_j',Y_j')$, the inclusion isotonicity property (Section 6.2) of the natural interval extension would imply that (9.3) holds for $(X_k,Y_k) = (X_{free},Y_{free})$. However, the implementation does not necessarily guarantee this property with mathematical rigor and, moreover, (9.2) was obtained in a slightly different coordinate range, (see Remark 8.10 and Section 8.4 for details). Thus (9.3) had to be explicitly verified using (X_{free},Y_{free}) for (X_k,Y_k).

Assertion 3 *For* $n = 28, 29, 30$, $(\bar{x},\bar{y})^r$ *is the only optimal point arrangement in* $(X,Y)^{*,r}$.

This assertion was proved by using the interval version of the theorem by Nurmela and Östergård, [88]:

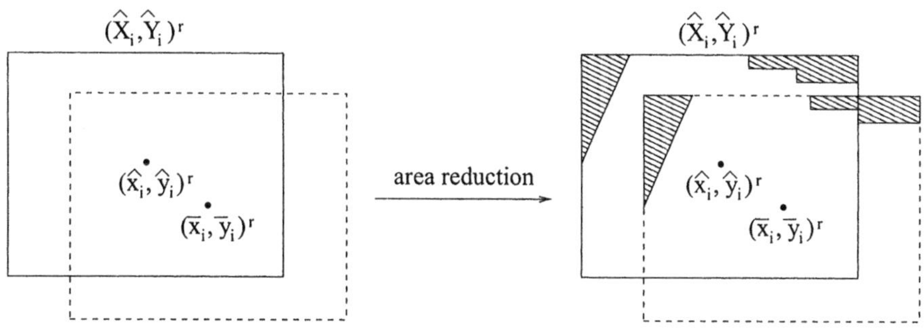

Fig. 9.1. Area reduction of the real (solid border) and hypothetical (dashed border) error rectangles in the theorem by Nurmela and Östergård.

Theorem 9.1. (based on [88]) *Assume that we know an approximation* $(\hat{x}, \hat{y})^r$ *of* $(\bar{x}, \bar{y})^r$*, such that*

$$\forall i \; \sqrt{d((\hat{x}_i, \hat{y}_i)^r, (\bar{x}_i, \bar{y}_i)^r))} < d' \tag{9.4}$$

with a suitable $d' \in \mathbb{R}$*. Determine an error rectangle* $(\hat{X}_i, \hat{Y}_i)^r \subseteq [0,1]^2$ *for each* i*, such that*

$$\forall i \; (\hat{x}_i, \hat{y}_i)^r \in (\hat{X}_i, \hat{Y}_i)^r \quad \text{and} \quad \forall i \; (X_i, Y_i)^{*,r} \subseteq (\hat{X}_i, \hat{Y}_i)^r, \tag{9.5}$$

and assume that the second statement of (9.5) also holds for an identical rectangle obtained by shifting the coordinates of the original rectangle with the vector $(\bar{x}_i - \hat{x}_i, \bar{y}_i - \hat{y}_i)$ *from* $(\bar{x}_i, \bar{y}_i)^r$*. Choose a proper real value (called as a cutoff value)* $\hat{f}$*, for which*

$$\hat{f} + 2d' < f^* \tag{9.6}$$

holds. Run an elimination algorithm on $(\hat{X}, \hat{Y})^r$ *which attempts to remove all points of* $(x, y)^r \in (\hat{X}, \hat{Y})^r$ *for which* $f(x, y) < \hat{f}$*. If we are able to achieve the contraction of each* $(\hat{X}_i, \hat{Y}_i)^r$ *component, then it is guaranteed that* $(\bar{x}, \bar{y})^r$ *is the only optimal point arrangement within* $(\hat{X}, \hat{Y})^r$ *and, thus within* $(X, Y)^{*,r}$*.*

The proof of the present theorem is much like the one in [88] and so it shall not be repeated here. The essence of the proof is that if some regions can be eliminated from the $(\hat{X}_i, \hat{Y}_i)^r$ error rectangles around $(\hat{x}_i, \hat{y}_i)^r$ using the $\hat{f}$ cutoff value, then congruent regions can be eliminated from the (hypothetical) error rectangles around $(\bar{x}_i, \bar{y}_i)^r$ using the f^* cutoff value (see Figure 9.1). That is, a contraction of the rectangles drawn around the approximate solution implies a contraction of the rectangles around the exact optimal solution.

To satisfy the conditions of Theorem 9.1, the following constructions were made: $(\hat{x}, \hat{y})^r$ was set to the existing approximate solution $(x_{appr}, y_{appr})^r$. Then the left-hand side of (9.4) was evaluated for each i using the interval enclosure of the arguments, resulting in a reliable d' bound. A mathematically correct $\hat{f}$ value was also evaluated in an interval way by using F^* and an enclosure of d' in (9.6). The error rectangles were computed to be $\hat{X}_i = (X_i^* + \Delta) \cap [0,1]$, $\hat{Y}_i = (Y_i^* + \Delta) \cap [0,1]$ with $\Delta =$

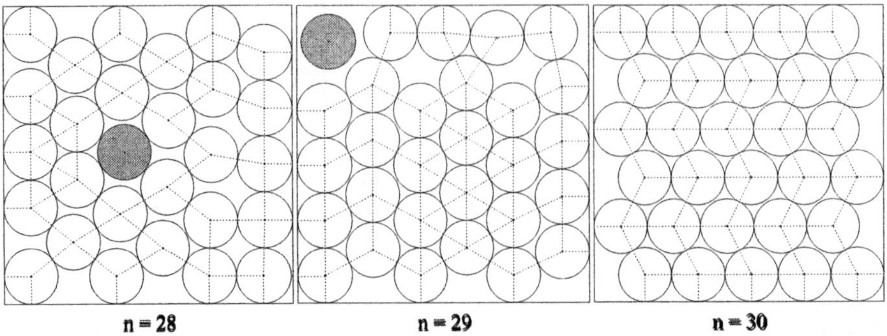

Fig. 9.2. The unique optimal packing structures of 28, 29, and 30 circles in a square.

$[-10^{-5}, 10^{-5}]$. This implied (9.5), and allowed us to verify the required property for the shifted rectangles as well—after evaluating an interval enclosure of the shift vector. Note that the choice of the order of Δ has a demonstrative role: since d' is of order 10^{-15}, the prerequisites for the theorem would still apply for much smaller error rectangles.

The required elimination algorithm of Theorem 9.1 was the advanced interval B&B optimization method of Chapter 8. As we saw, this algorithm provides a guarantee that only packings with $f(x, y) < \hat{f}$ are eliminated from $(\hat{X}, \hat{Y})^r$. The algorithm was able to verify the required contraction property for each component without any particular difficulties (within 100 iteration steps), and consequently, to prove Assertion 3 for each of the three packing problems.

Corollary 9.2. *The packing structures conjectured in [36] for $n = 28$ and in [86] for $n = 29$ and 30 result in the only optimal packings of n points in the unit square (disregarding symmetric cases and the movement of the free point for $n = 28, 29$). Hence,* the conjectured optimal structure may be regarded as the unique optimal geometric solution for the given packing problem.

PROOF. Let $f((\bar{x}, \bar{y})^r) = \bar{f}$, and let $(x', y') \subseteq (X, Y)^*$, $(x', y')^r \neq (\bar{x}, \bar{y})^r$ be an arbitrary point arrangement of n points. Then by Assertion 3, $f((x', y')^r) < \bar{f}$, that is, $f(x', y') < \bar{f}$. Since by Assertion 2 $(\bar{x}, \bar{y})^r$ can be extended to $(\bar{x}, \bar{y})$ such that $f(\bar{x}, \bar{y}) = \bar{f}$, we obtain the result that $\bar{f} = f^*$. Thus $(\bar{x}, \bar{y})$ is optimal and unique (apart from a symmetry factor and a possible free circle), and by Assertion 1 this is determined from the conjectured packing structure. □

Figure 9.2 shows the optimal packing structures for 28, 29, and 30 circles (points) in the square, respectively, showing the free and touching circles as well.

10

Repeated Patterns in Circle Packings

On comparing the structures of the optimal and the best known circle packings, we have noticed some repeated patterns [36, 86]. In some cases there is a connection between the given structure and the number of circles. Based on this fact, we can classify some packings into pattern classes.

10.1 Finite pattern classes

We use the notation **PAT**$(f(k))$ for the pattern classes and its patterns, where $f : \mathbb{N} \to \mathbb{N}$ is a function and $f(k)$ is the number of circles ($n = f(k)$).

10.1.1 The PAT$(k^2 - l)$ $(l = 0, 1, 2)$ pattern classes

PAT(k^2): Perhaps the most conspicuous pattern belongs to the $n = 4, 9, 16, 25$ and 36 square numbers, when the circles have a $k \times k$ ($n = k^2$) quadratical structure (Figure 10.1). In these circle packings, the coordinates of points of the related point arrangements are (il_k, jl_k), where $0 \leq i, j \leq k-1$ are integer numbers, and $l_k = 1/(k-1)$.

Assertion 4 *In the point arrangements using pattern* **PAT**(k^2), $(k \geq 2)$, *the minimal distance between the points is*

$$m_n = l_k = \frac{1}{k-1}.$$

PROOF. The distance between the $(i_1 l_k, j_1 l_k)$ and $(i_2 l_k, j_2 l_k)$ points is

$$\frac{\sqrt{(i_1 - i_2)^2 + (j_1 - j_2)^2}}{k-1} \geq l_k,$$

because at least one square of differences in the numerator is not zero. The equality is realized if and only if, exactly one is fulfilled from the equations $|i_1 - i_2| = 1$ and

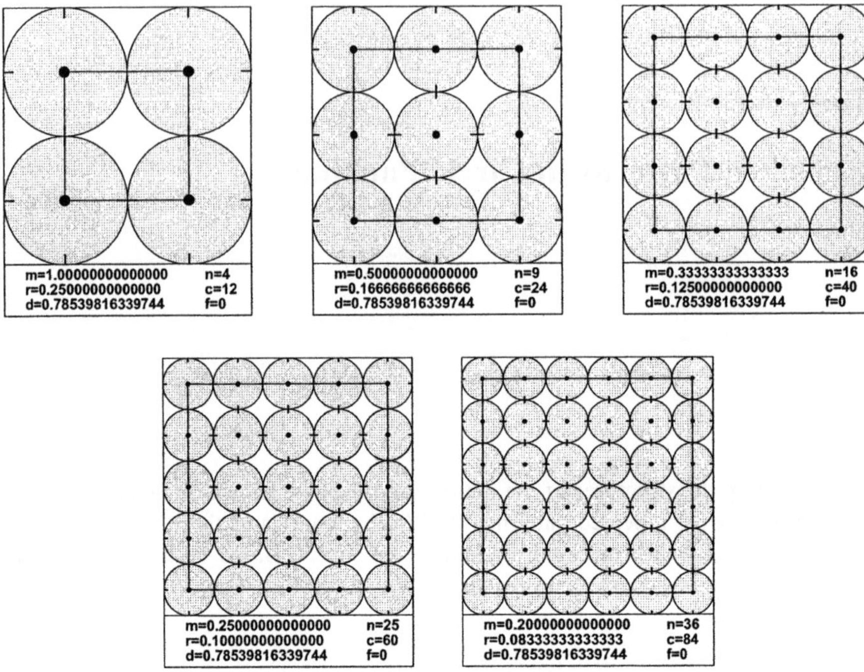

Fig. 10.1. Circle packings in the pattern class **PAT**(k^2), where $k = 2, 3, 4, 5$, and 6.

$|j_1 - j_2| = 1$. □

It has been proved that this pattern gives the optimal circle packings for $k = 2, 3, 4$ and 5 (and probably also for $k = 6$), but it is certain that for $k = 7$ it is not optimal. For the point packing problem with $n = 49$ the maximal minimal pairwise distance value m_{49} is greater than 0.1673 (see in the Appendix), but using the **PAT**(k^2) pattern for $k = 7$, the minimal distance between the points is $\frac{1}{6} \approx 0.1667$. It is interesting to note that the density of the circle packings in this pattern class is always equal to a constant, because

$$d_n(r_n, [0,1]^2) = k^2 \left(\frac{1}{2k}\right)^2 \pi = \frac{\pi}{4}.$$

PAT$(k^2 - 1)$: Locate $n = k^2$ circles in the square ($k \geq 3$) based on the pattern **PAT**(k^2), and remove a circle from the packing that has no connection with the sides of the square. The remaining $k^2 - 1$ circles can be located in a smaller side square after we shift the circles in the selected column (and row) in parallel to the side (0,0)-(0,1) (and respectively, in parallel with the side (0,0)-(1,0)), we compress the full circle packing in such a way that the centre of the circles that is connected by the chosen circle generates a rhombus.

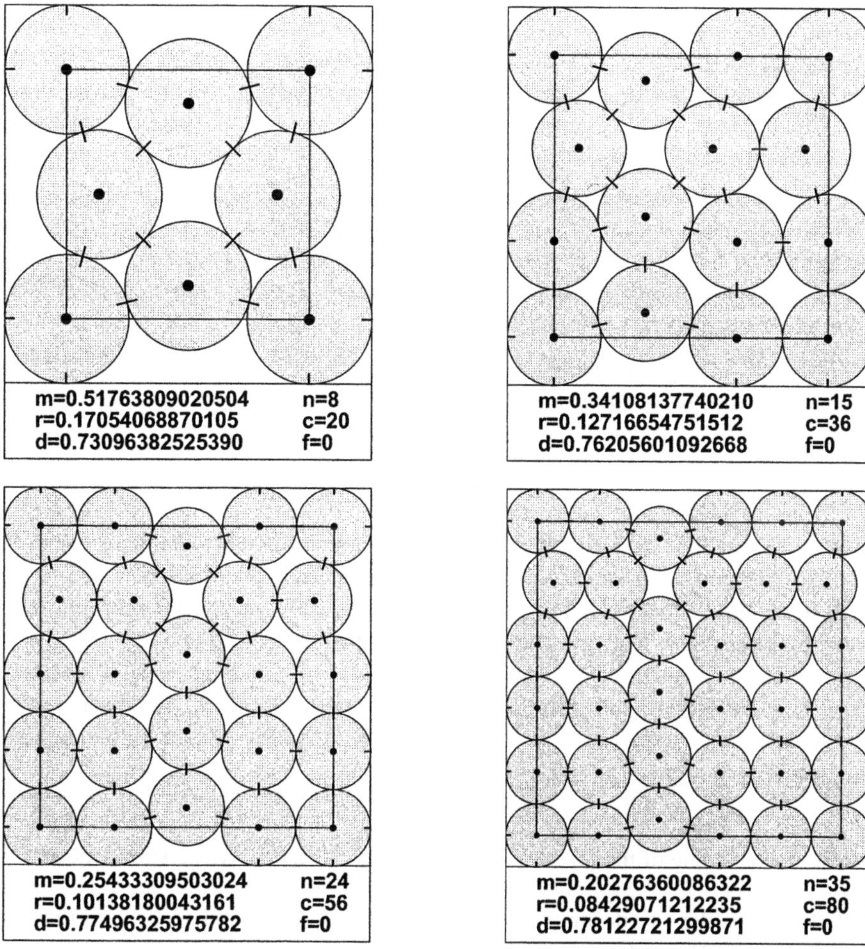

Fig. 10.2. Circle packings in the pattern class of **PAT**(k^2-1), where $k = 3, 4, 5$, and 6.

In the equivalent point arrangement problem, if we remove that circle centre which has the $\left(\frac{k-2}{k-1}, \frac{k-2}{k-1}\right)$ coordinates, then the coordinates of the remaining points will be $\bigcup_{i=1}^{6} C_i$, where

$$l_k = \frac{1}{k-3+\sqrt{2+\sqrt{3}}}$$

and

$$\begin{aligned}
C_1 &= \{(il_k, jl_k) \mid 0 \le i, j \le k-3\}\,,\\
C_2 &= \{(1, il_k) \mid 0 \le i \le k-3\}\,,\\
C_3 &= \{(il_k, 1) \mid 0 \le i \le k-3\}\,,\\
C_4 &= \{((i + \sin 15^o)l_k, (k-2+\cos 15^o)l_k) \mid 0 \le i \le k-3\}\,,\\
C_5 &= \{((k-2+\cos 15^o)l_k, (i + \sin 15^o)l_k) \mid 0 \le i \le k-3\}\,,\\
C_6 &= \{(1,1), (1 - l_k \sin 15^o, (k-2+\cos 15^o)l_k),\\
&\quad ((k-2+\cos 15^o)l_k, 1 - l_k \sin 15^o)\}.
\end{aligned}$$

Assertion 5 *In the point arrangement problems defined by* **PAT**(k^2-1)*,* $k \ge 3$ *the minimal distance between the points is*

$$m_n = l_k = \frac{1}{k - 3 + \sqrt{2+\sqrt{3}}}.$$

PROOF. Let us first show that the distance between two points taken from a set C_i $(1 \le i \le 6)$, and the distance between two different sets C_i and C_j $(1 \le i, j \le 6)$ is greater than or equal to l_k. This requires considering 33 cases, from which we will prove two cases here. The other proofs are quite similar.

a) Let us consider the $(i_1 l_k, j_1 l_k)$ and $(i_2 l_k, j_2 l_k)$ points from C_1. The distance between them is

$$\sqrt{(i_1 - i_2)^2 l_k^2 + (j_1 - j_2)^2 l_k^2} \ge \sqrt{l_k^2} = l_k,$$

because at least one of the $(i_1 - i_2)^2$ and $(j_1 - j_2)^2$ is a positive square number.

b) Let us consider the distance between $(i_1 l_k, j_1 l_k) \in C_1$ and $((i_2 + \sin 15^o)l_k,$ $k - 2 + \cos 15^o)l_k) \in C_4$, which is

$$l_k \sqrt{(i_2 + \sin 15^o - i_1)^2 + (k - 2 + \cos 15^o - j_1)^2}.$$

We have to show that the sum in the square root is greater than or equal to 1. It is easy to see that the second part of the sum is always greater than 1 if $j_1 \ne k-2$, and the first part is also greater than 1 if $i_2 > i_1$.

Let us suppose that $j_1 = k-2$ and $i_2 = i_1$. In this case the values in the square root is 1. Let us consider the case when $j_1 = k-2$ and $i_2 < i_1$. Here the sum under the square root is

$$1 + (i_1 - i_2)(i_1 - i_2 - 2\sin 15^o),$$

which is obviously greater than 1, because $0 < 2\sin 15^o < 1$. □

It can be proved that these circle packings for the $k = 3, 4,$ and 5 cases are optimal (Figure 10.2). This pattern probably provides an optimal circle packing for the $k = 6$ $(n = 35)$ case, but not for the $n = 48$ case. There is a point arrangement where m_{48}

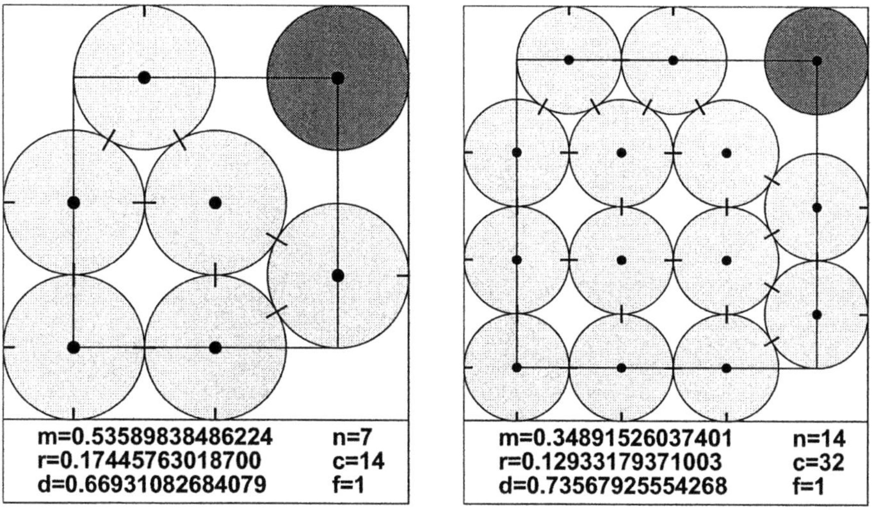

Fig. 10.3. Circle packings in the pattern class of $\mathbf{PAT}_1(k^2-2)$, where $k=3$ and 4.

is greater than 0.1694 (see it in the Appendix), but using the previous pattern, the minimal distance between the points is just

$$\frac{1}{4+\sqrt{2+\sqrt{3}}} \approx 0.1685.$$

In the case of $n=16$, owing to symmetry reasons, there is only one circle that we can remove from the square. But for $n=25$ there are two different such circles in the packing. These give optimal packings with different, but quite similar structures. For $n=36$, we have three such kind of circles.

PAT(k^2-2) From an analogy of the previous pattern, from the **PAT**(k^2) circle packings we can remove two circles together. Here we distinguish two subclasses.

a) $\mathbf{PAT}_1(k^2-2)$ Consider a circle in a packing of the pattern class **PAT**(k^2) that is located in a corner of the square. Remove those two circles that touch the selected one. In this way, both in the row and in the column of the selected circle, $k-1$ circles remain, and the chosen one becomes a free circle.

The $2\times(k-1)$ circles in the row, and in the column of the chosen circle can be placed inside the circles on the respective sides of the square packing of $(k-1)\times(k-1)$ circles in such a way that we match one new circle to two neighbouring ones on the side of the square packing. The packing obtained (together with the free circle in the corner) can be squeezed into a square smaller than the earlier one. The coordinates of the respective point packing are, if we have removed the circles corresponding to the points $\left(\frac{k-2}{k-1},1\right)$ and $\left(1,\frac{k-2}{k-1}\right)$:

$$\bigcup_{i=1}^{4} C_i,$$

where

$$l_k = \frac{1}{k - 2 + \frac{\sqrt{3}}{2}},$$

and

$$C_1 = \{(il_k, jl_k) \mid 0 \leq i, j, \leq k - 2\},$$
$$C_2 = \left\{(\frac{l_k}{2} + il_k, 1) \mid 0 \leq i \leq k - 3\right\},$$
$$C_3 = \left\{(1, \frac{l_k}{2} + il_k) \mid 0 \leq i \leq k - 3\right\},$$
$$C_4 = \{(1, 1)\} \text{ (free circle).}$$

Assertion 6 *In the point arrangements using pattern* $\mathbf{PAT}_1(k^2 - 2)$, $k \geq 3$, *the minimal distance between the points is*

$$m_n = l_k = \frac{1}{k - 2 + \frac{\sqrt{3}}{2}}.$$

PROOF. Here, again, we just check a few critical cases. The others can be completed in a similar way.

a) Let $(\frac{l_k}{2} + i_1 l_k, 1)$ and $(\frac{l_k}{2} + i_2 l_k, 1)$ be elements of C_2. Since $i_1 \neq i_2$, the distance $\sqrt{(i_1 - i_2)^2 l_k^2}$ between them is obviously not less than l_k.

b) Consider the distance between the points $(i_1 l_k, j_1 l_k)$ (of C_1) and $(\frac{l_k}{2} + i_2 l_k, 1)$ (of C_2):

$$\sqrt{\left(\frac{l_k}{2} + i_2 l_k - i_1 l_k\right)^2 + (1 - j_1 l_k)^2}.$$

In the case when $j_1 \neq k - 2$ the second term in the square root is larger than l_k^2, and so it is enough to consider the case of $j_1 = k - 2$. In a similar way if $i_1 \neq i_2$ and $i_1 - i_2 \neq 1$, then the first term of the sum is larger than l_k^2. If $i_1 = i_2$ or $i_1 - i_2 = 1$ and $j_1 = k - 2$, then the distance between the two points is exactly l_k. □

For $k = 3$ and 4 the $\mathbf{PAT}_1(k^2 - 2)$ pattern provides the optimal packings (Figure 10.3). With $k = 5$ and 6 the structure of the optimal and the best known packings (Figure 10.4) can be obtained in a similar way. The only difference is that the two circles are taken out from the inside of the **PAT**(k^2) packing (earlier they were deleted from a side of the packing).

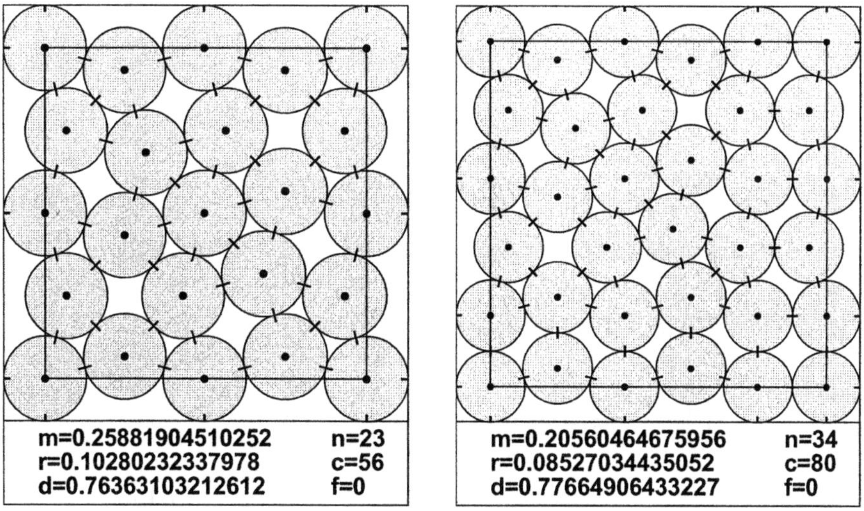

Fig. 10.4. Circle packings in the pattern class of $\mathbf{PAT}_2(k^2-2)$, where $k=5$ and 6.

b) $\mathbf{PAT}_2(k^2-2)$ Consider a packing of $\mathbf{PAT}(k^2)$ for which $k \geq 5$. Take out the circles corresponding to the points $\left(\frac{k-2}{k-1}, \frac{k-2}{k-1}\right)$ and $\left(\frac{k-4}{k-1}, \frac{k-4}{k-1}\right)$, and push the remaining ones in the direction of the earlier, removed circles. We can now compress the packing along the sides of the square. The centres of the four circles that touched earlier the removed ones form a square. The coordinates of the corresponding point arrangement are: $\bigcup_{i=1}^{6} C_i$, where

$$l_k = \frac{1}{k-5+2\sqrt{2+\sqrt{3}}}$$

and

$$\begin{aligned}
C_1 &= \{(il_k, jl_k) \mid 0 \leq i, j, \leq k-5\}, \\
C_2 &= \{(l_k(k-5)+2il_k\cos 15^o, jl_k) \mid 1 \leq i \leq 2;\ 0 \leq j \leq k-5\}, \\
C_3 &= \{(jl_k, l_k(k-5)+2il_k\cos 15^o) \mid 1 \leq i \leq 2;\ 0 \leq j \leq k-5\}, \\
C_4 &= \{(l_k(k-5)+il_k\cos 15^o, l_k\sin 15^o+jl_k) \mid 1 \leq i \leq 2;\ 0 \leq j \leq k-5\}, \\
C_5 &= \{(l_k\sin 15^o+jl_k, l_k(k-5)+il_k\cos 15^o) \mid 1 \leq i \leq 2;\ 0 \leq j \leq k-5\}, \\
C_6 &= \{(1-il_k\cos 15^o, 1-(i \bmod 2)l_k\sin 15^o) \mid 1 \leq i \leq 3\}, \\
C_7 &= \{(1-(i \bmod 2)l_k\sin 15^o, 1-il_k\cos 15^o) \mid 1 \leq i \leq 3\},
\end{aligned}$$

$$\begin{aligned}C_8 = \{&(l_k(k-5) + 2l_k \cos 15^o, l_k(k-5)\cos 15^o + 2l_k \cos 15^o),\\ &\left(l_k(k-5) + l_k \cos 15^o, l_k \sin 15^o + (k-5)l_k + \sqrt{2}l_k\right),\\ &\left(1 - l_k \sin 75^o, 1 - l_k \sin 15^o - \sqrt{2}l_k\right),\\ &\left(1 - l_k \sin 15^o - \sqrt{2}l_k, 1 - l_k \sin 75^o\right),\\ &\left(l_k \sin 15^o + (k-5)l_k + \sqrt{2}l_k, l_k(k-5) + l_k \sin 75^o\right),\\ &\left(l_k(k-5) + \frac{\sqrt{6}}{2}l_k, 1 - \frac{\sqrt{6}}{2}l_k\right),\\ &\left(1 - \frac{\sqrt{6}}{2}l_k, l_k(k-5) + \frac{\sqrt{6}}{2}l_k\right), (1,1)\}.\end{aligned}$$

Assertion 7 *For point arrangement problems of the pattern class* $\mathbf{PAT}_2(k^2-2)$ *with* $k \geq 5$ *the equality*

$$m_n = l_k = \frac{1}{k - 5 + 2\sqrt{2 + \sqrt{3}}}$$

holds.

PROOF. Out of the 98 cases to be checked here we consider only two, the remaining ones can be proved in a similar way.

a) Let $(l_k(k-5) + 2i_1 l_k \cos 15^o, j_1 l_k)$ and $(l_k(k-5) + 2i_2 l_k \cos 15^o, j_2 l_k)$ be in C_2. The distance between the two points is

$$\sqrt{4l_k^2 \cos^2 15^o (i_1 - i_2)^2 + l_k^2 (j_1 - j_2)^2}.$$

When $i_1 \neq i_2$ the first term of the sum in the square root is larger than l_k^2 since $\cos 15^o > 0.9$. It is easy to show in a similar way that the second term of the sum is larger than l_k^2 when $j_1 \neq j_2$. As we have considered two separate points in C_2, at least one of the two terms must be non-zero.

b) Assume now that the point $(i_1 l_k, j_1 l_k)$ is in C_1 and $(l_k(k-5) + i_2 l_k \cos 15^o, l_k \sin 15^o + j_2 l_k)$ is in C_4. Then the distance between the two is

$$\sqrt{(l_k(k-5) + i_2 l_k \cos 15^o - i_1 l_k)^2 + (l_k \sin 15^o + j_2 l_k - j_1 l_k)^2}.$$

Due to the fact that $0 \leq i_1 \leq k-5$ and $1 \leq i_2 \leq 2$, the first term of the sum in the square root is never less than $l_k^2 \cos^2 15^o$.

It is obvious that the second term cannot be less than $l_k^2 \sin^2 15^o$ either, when $j_2 \geq j_1$. This is also the case when $j_2 < j_1$, since after introducing the notion of $u = j_1 - j_2$ the second term will become $l_k^2(\sin 15^o - u)^2$, where the second part is

$$\sin^2 15^o + u^2 - 2u \sin 15^o > \sin^2 15^o,$$

as $u \geq 1 > 2 \sin 15^o$. □

For $k = 5$ the **PAT**$_2(k^2 - 2)$ pattern provides the optimal circle packing, and for $k = 6$ the pattern is the best known packing. In the case of $k = 7$, i.e. for $n = 47$ circles the TAMSASS-PECS algorithm discussed in Section 4.2 was able to find a better packing that belongs to the **PAT**$_2(k^2 - 2)$ pattern (see the Appendix: m_{47} is greater than 0.1712, while $\frac{1}{2+2\sqrt{2+\sqrt{3}}} \approx 0.1705$).

It is an interesting question whether one can find similar repeated patterns for $n = k^2 - l$ circles, when $l \geq 3$. The answer is that apparently we can have certain similar structures, but these are so complex that it is a very challenging problem to give the exact positions of the circles. Nobody has been able to supply such repeated packing patterns so far. In the next subsection we shall not elaborate on patterns; only approximations are given in the form of suitable structure classes [114].

10.1.2 The STR($k^2 - l$) ($l = 3, 4, 5$) structure classes

The notation **STR**($f(k)$) will be used in practically the same way as **PAT**($f(k)$), i.e. $f : N \to N$ denotes a function, and $f(k)$ tells us the number of circles ($n = f(k)$) [114]. The modified notation (from pattern class to structure class) is based on the argument that this time the coordinates of the optimal, and the best known packings are not given, just a well approximating structure is defined.

When we study the optimal and best known circle packings for $n = k^2 - 3$ ($k = 4, 5, 6, 7, 8$) we find that they have similar structures (see Figure 10.5).

These packings are quite similar to these cases when we remove from the **PAT**(k^2) pattern the circles belonging to the points

$$(1,0), \left(\frac{k-2}{k-1}, \frac{2}{k-1}\right)$$

and

$$\left(\frac{k-3}{k-1}, \frac{1}{k-1}\right),$$

and we compress the packing as much as possible. The structure of the packings obtained using this earlier idea is not the same, but they are quite similar. Let us consider these structure classes one by one.

STR($k^2 - 3$) Consider those circle packings that correspond to the point arrangements $\bigcup_{i=1}^{6} C_i$, where $l_k = \frac{1}{k-3+\sqrt{3}}$ and

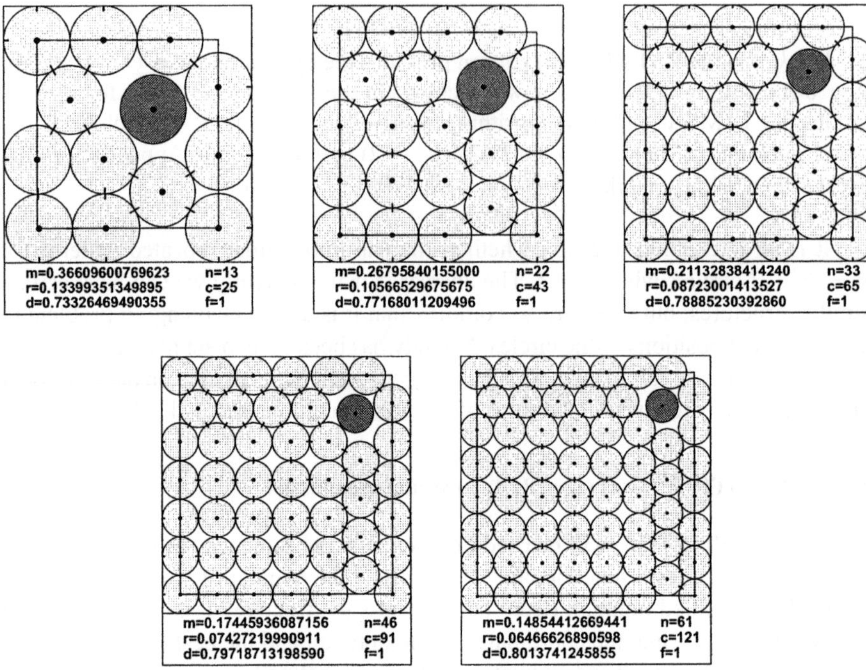

Fig. 10.5. Circle packings of the structure class **STR**(k^2-3), where $k = 4, 5, 6, 7$, and 8.

$$\begin{aligned}
C_1 &= \{(il_k, jl_k)|\ 0 \le i, j \le k-3\}\,,\\
C_2 &= \{(1, il_k)|\ 0 \le i \le k-2\}\,,\\
C_3 &= \{(il_k, 1)|\ 0 \le i \le k-2\}\,,\\
C_4 &= \left\{(1-\frac{\sqrt{3}l_k}{2}, \frac{l_k}{2}+il_k)|\ 0 \le i \le k-4\right\},\\
C_5 &= \left\{(\frac{l_k}{2}+il_k, 1-\frac{\sqrt{3}}{2}l_k)|\ 0 \le i \le k-4\right\},\\
C_6 &= \{(1-l_k \sin 75^o, 1-l_k \sin 75^o)\} \quad \text{(free circle)}.
\end{aligned}$$

Assertion 8 *For the point arrangements of the structure class* **STR**(k^2-3), *the following relations hold for all* $k \ge 4$*:*

$$m_n = l_k = \frac{1}{k-3+\sqrt{3}} \quad (\le \overline{m}_n).$$

PROOF. The checking of the 20 possible cases can be carried out in a similar way as before. □

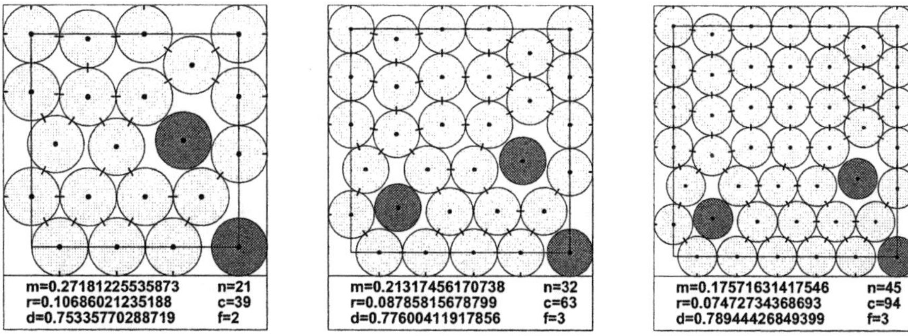

Fig. 10.6. Circle packings of the structure class **STR**($k^2 - 4$), where $k = 5, 6,$ and 7.

STR($k^2 - 4$) The structural similarity of the optimal and best known circle packings for $n = k^2 - 4$ ($k = 5, 6, 7$) is obvious too (see Figure 10.6).

A good approximation of these packings can be obtained when we place the circles according to the pattern **PAT**(k^2), and we remove those circles that are related to the points

$$\left(\frac{k-2}{k-1}, 0\right), \left(1, \frac{1}{k-1}\right), \left(\frac{k-2}{k-1}, \frac{3}{k-1}\right),$$

and

$$\left(\frac{k-3}{k-1}, \frac{2}{k-1}\right)$$

of the respective point arrangement problems. The packing is also compressed, as before. The coordinates of the resulting point arrangements will be $\bigcup_{i=1}^{7} C_i$, where $l_k = \frac{1}{k-3+\sqrt{3}}$,

$$C_1 = \{(il_k, 1 - jl_k) |\ 0 \le i \le k-3, 0 \le j \le k-4\},$$

$$C_2 = \{(1, 1 - jl_k) |\ 0 \le j \le k-3\},$$

$$C_3 = \left\{(1 - \frac{\sqrt{3}}{2} l_k, 1 - \frac{2j+1}{2} l_k) |\ 0 \le j \le k-5\right\},$$

$$C_4 = \left\{(\frac{2i+1}{2} l_k, 1 - (k - 4 - \frac{\sqrt{3}}{2}) l_k) |\ 0 \le i \le k-4\right\},$$

$$C_5 = \left\{(il_k, \frac{\sqrt{3}}{2} l_k) |\ 0 \le i \le k-2\right\},$$

$$C_6 = \left\{(\frac{1+2i}{2} l_k, 0) |\ 0 \le i \le k-3\right\},$$

$$C_7 = \left\{(1, 0), ((k-3) l_k + \frac{\sqrt{2}}{2} l_k, 1 - (k-4) l_k + \frac{\sqrt{2}}{2} l_k)\right\} \quad \text{(free circles).}$$

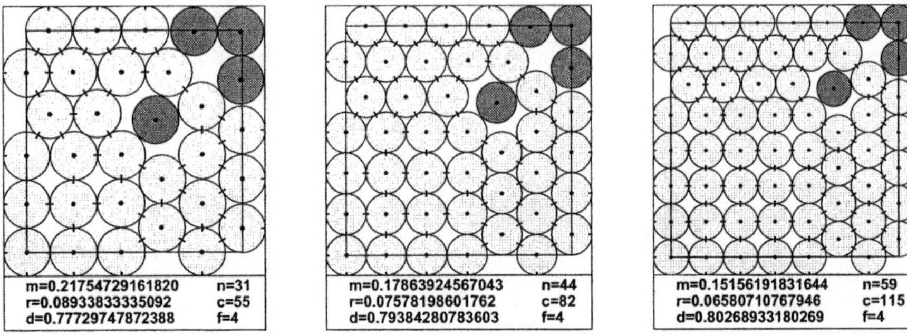

Fig. 10.7. Circle packings of the structure class **STR**(k^2-5), where $k=6,7$, and 8.

Assertion 9 *For the point arrangement problem cases of the* **STR**(k^2-4) *structure class for all* $k \geq 5$ *the relations*

$$m_n = l_k = \frac{1}{k-3+\sqrt{3}} \; (\leq \overline{m}_n).$$

hold.

PROOF. The 34 possible cases can be proved in a similar way as before. □

STR(k^2-5) The circle packing cases of $n=k^2-5$ have similar structures as above for $k=6,7,8$ (see Figure 10.7).

We can again approximate the packings if the structure is generated in the following way. Take the points of the point arrangement belonging to a circle packing class of **PAT**(k^2). Remove the points with the coordinates of

$$\left(\frac{k-3}{k-1},1\right),\left(\frac{k-2}{k-1},\frac{k-2}{k-1}\right),\left(1,\frac{k-3}{k-1}\right),\left(\frac{k-4}{k-1},\frac{k-3}{k-1}\right),\left(\frac{k-3}{k-1},\frac{k-4}{k-1}\right),$$

and compress the structure obtained as much as possible. These packings contain four free circles and the packings are symmetric for the (0,0)–(1,1) diagonal of the square. Now the coordinates of the point arrangements can be specified as $\bigcup_{i=1}^{7} C_i$, where

$$l_k = \frac{1}{k-4+\frac{3}{2}\sqrt{3}},$$

and

$$C_1 = \left\{(il_k, jl_k) \mid 0 \leq i,j \leq k-4\right\},$$

$$C_2 = \left\{(\frac{2i+1}{2}l_k, (k-4+\frac{\sqrt{3}}{2})l_k) \mid 0 \leq i \leq k-5\right\},$$

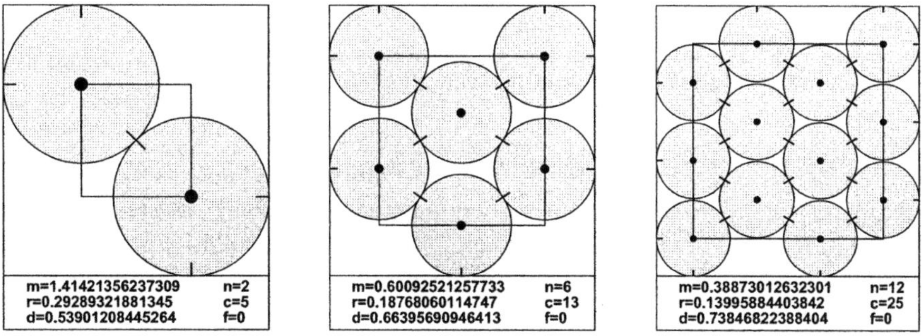

Fig. 10.8. Circle packings belonging to the $\mathbf{PAT}_1(k(k+1))$ class, for $k = 1, 2$, and 3.

$$C_3 = \left\{ ((k-4+\frac{\sqrt{3}}{2})l_k, \frac{2j+1}{2}l_k) |\, 0 \le j \le k-5 \right\},$$

$$C_4 = \left\{ (il_k, (k-4+\sqrt{3})l_k) |\, 0 \le i \le k-3 \right\},$$

$$C_5 = \left\{ ((k-4+\sqrt{3})l_k, jl_k) |\, 0 \le j \le k-3 \right\},$$

$$C_6 = \left\{ (\frac{2i+1}{2}l_k, 1), 0 \le i \le k-4 \right\},$$

$$C_7 = \left\{ (1, \frac{2i+1}{2}l_k), 0 \le i \le k-4 \right\},$$

$$C_8 = \left\{ (1,1), (1-l_k, 1), (1, 1-l_k), ((k-4+\frac{\sqrt{2}}{2})l_k, (k-4+\frac{\sqrt{2}}{2})l_k) \right\}$$

(the latter are free circles).

Assertion 10 *For all the point arrangement problems belonging to the* $\mathbf{STR}(k^2-5)$ *structure class with* $k \ge 6$ *the relations*

$$m_n = l_k = \frac{1}{k-4+\frac{3}{2}\sqrt{3}} \; (\le \overline{m}_n)$$

hold.

PROOF. The proof for the 62 possible subcases is similar to what we did for the last assertions. □

Repeated patterns are not necessarily related to square-like $\mathbf{PAT}(k^2)$ structures. There also exist other grid packing patterns among the optimal and best known packings.

10.1.3 The PAT($k(k+1)$) pattern class

Two distinct subclasses exist within this pattern class:

PAT$_1(k(k+1))$ Place points on a rectangular grid in the following way. Subdivide one side of the square into k parts, another side perpendicular into the previous one to $2k-1$ equal parts. Next, draw lines through the intersection points parallel to the sides. Place one point in every second grid point, starting with a square having coordinates (0,0)—see Figure 10.8. Then the coordinates of the related grid packing can be given as

$$\left(\frac{i}{k}, \frac{i \bmod 2}{2k-1} + \frac{2j}{2k-1}\right),$$

where $0 \leq i \leq k$ and $0 \leq j \leq k-1$.

It is easy to see that the m_n minimal distance between the points must be equal to

$$\sqrt{\frac{1}{k^2} + \frac{1}{(2k-1)^2}},$$

which is exactly the length of the diagonal of a small rectangle if $1 \leq k \leq 3$.

Although in these cases the optimal packings belong to this pattern, for larger k values the optimal and best known packings form different pattern classes.

PAT$_2(k(k+1))$ Consider now the point arrangements defined by the coordinates

$$\left(i\sqrt{l_k^2 - (1 - kl_k + l_k)^2}, i \bmod 2 + (-1)^i j l_k\right),$$

where $0 \leq i \leq k,\ 0 \leq j \leq k-1$, $k \geq 4$,

$$l_k = \frac{k^2 - k - \sqrt{2k}}{k^3 - 2k^2}.$$

Then the minimal distance between the points is $m_n = l_k$. The optimal packing belongs to this pattern class when $k = 4$ and 5, and they are the best known packings for the cases $k = 6$ and 7 (see Figure 10.9).

10.1.4 The PAT($k^2 + \lfloor k/2 \rfloor$) pattern class

Similar to the construction of the **PAT**$_1(k(k+1))$ pattern, consider point arrangements when the sides of the square are subdivided into k and $2k-2$ equal parts (in contrast to the earlier procedure with k and $2k-1$ parts, respectively).

Then the point arrangements can be given by the coordinates

$$\left(\frac{i}{k}, \frac{j}{2k-2}\right),$$

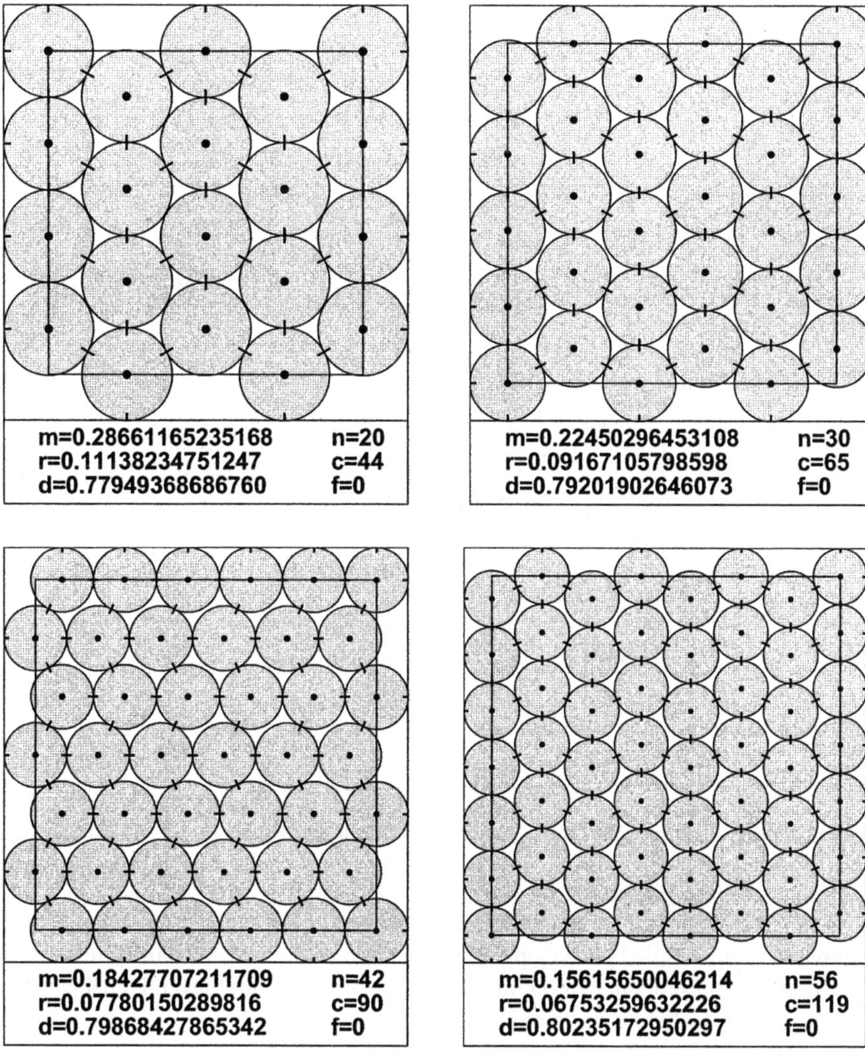

Fig. 10.9. Circle packings from the pattern class $\mathbf{PAT}_2(k(k+1))$ for $k = 4, 5, 6$, and 7.

where $0 \le i \le k, 0 \le j \le 2k - 2$ and $k \le 7$.

Then

$$m_n = \sqrt{\frac{1}{k^2} + \frac{1}{(2k-2)^2}}.$$

However, the minimal distance between the points can only be equal to the length of the diagonal of the small rectangles when $k^2 - 8k + 4 \le 0$. This can be readily

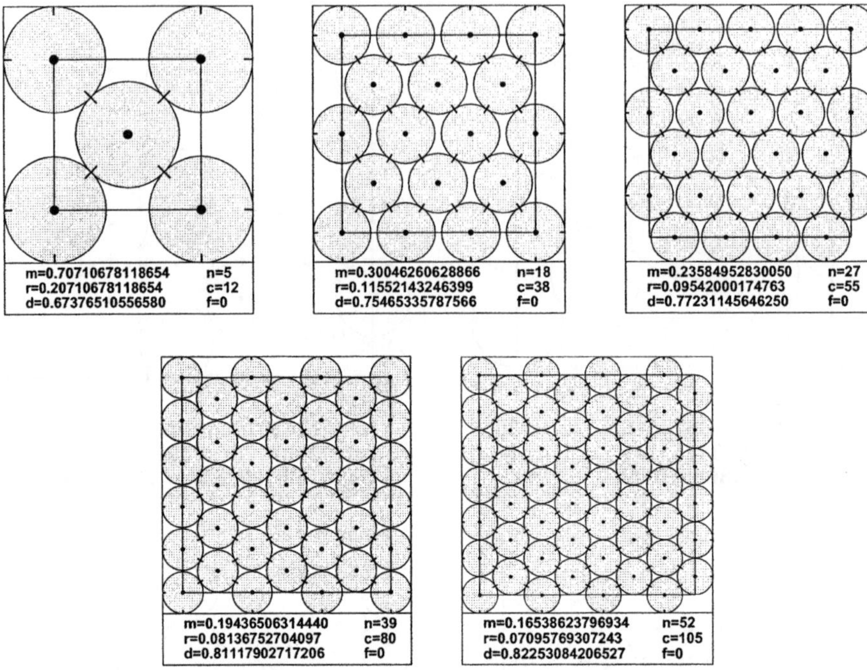

Fig. 10.10. Circle packings of the **PAT**$(k^2 + \frac{k}{2})$ pattern class, where $k = 2, 4, 5, 6$, and 7.

proved, in this case $k \leq 7$, where k is a positive integer. In the $k = 2, 4$ and 5 cases the optimal, for the $k = 6$ and 7 cases the best presently known packings are defined by this pattern (see Figure 10.10).

The patterns and structures we have introduced obviously cannot cover all the optimal structures. Still, by developing optimization algorithms and hardware we can learn more and more about new optimal packings, and in this way several other pattern classes will probably be discovered in the future.

10.2 A conjectured infinite pattern class

The patterns classes discussed up to this point do have the common property that all of them are optimal (or corresponding to the best presently known packing) only in a finite number of cases. Human intuition suggests that it is right in this way, since we expect that with an increasing number of circles to be packed the optimal packings will contain larger and larger hexagonal substructures. This means that the earlier patterns can provide the best packings only for a certain number of circles, then different structures should appear. Such a change in structure can be seen when we study the packings of $n = k(k + 1)$ circles. For the $k = 1, 2$ and 3 cases the pattern class $\mathbf{PAT}_1(k(k + 1))$ contains the optimal ones, and for increasing k values,

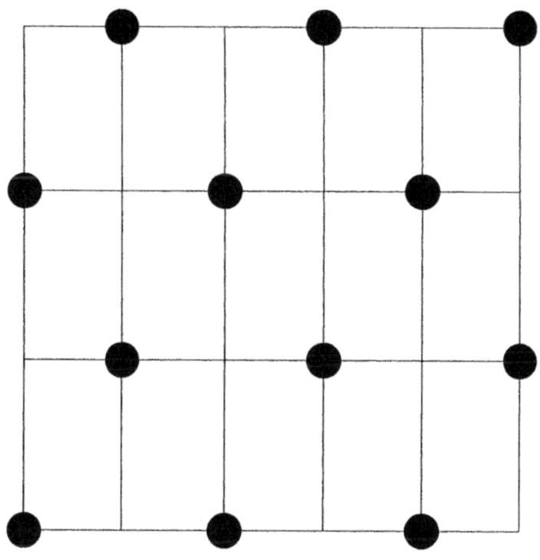

Fig. 10.11. The grid packing that is related to the $\frac{3}{5}$ fraction.

even for $k = 4, 5, 6$, and 7 the pattern class $\mathbf{PAT}_2(k(k+1))$ will provide the best packings.

In spite of the above line of thoughts, some experts think that an infinite pattern class may exist. K. J. Nurmela and his coworkers published the following conjecture in 1999 [89]: consider that subsequence of the approximating fraction sequence that is established when $\frac{\sqrt{3}}{3}$ is transformed to a simple continued fraction, which contains every second member of the original sequence. Then the periodic continued fraction

$$\frac{\sqrt{3}}{3} = [0; 1, 1, 2, 1, 2, 1, 2, 1, ...]$$

is

$$\frac{0}{1}, \frac{1}{1}, \frac{1}{2}, \frac{3}{5}, \frac{4}{7}, \frac{11}{19}, \frac{15}{26}, \frac{41}{71}, \ldots.$$

Now take the respective subsequence,

$$\frac{1}{1}, \frac{3}{5}, \frac{11}{19}, \frac{41}{71}, \ldots.$$

For each $\frac{p}{q}$ fraction of this sequence we can determine a point arrangement in the following way. Subdivide two perpendicular sides of the square into p and q parts, respectively. Draw lines parallel to the respective sides through the subdivision points (Figure 10.11). In this way we obtain $p \times q$ small rectangles. Place one point in the lower left corner of the square (in the (0, 0) point), and then place a point in every second grid point as we did before. It is not hard to see that in this way we can have

$$\left\lceil \frac{(p+1)(q+1)}{2} \right\rceil$$

points. Now we conjecture that each point arrangement formed in this way is based on a fraction of the previous fraction sequence and will be optimal.

This conjecture urges us to investigate these types of packing in more detail. Such packings will thenceforth be known as grid packings.

10.2.1 Grid packings

It is worth mentioning here that the grid packings have an important property: the minimal distance between the points can easily be computed: it is the common length of the diagonal of the grid rectangles. In cases when p and q are arbitrary natural numbers, this property does not hold for the related point arrangements. As can be easily seen, the packing has the above-mentioned property if and only if

$$\frac{p}{q} \in \left(\frac{\sqrt{3}}{3}, \sqrt{3} \right).$$

Here it is sufficient to consider an open interval from both sides, since the endpoints of the interval are irrational numbers. From now we shall let $[[p,q]]$ stand for those grid packings where p and q are integers such that $\frac{p}{q}$ fit into the above interval, and GP shall denote the set of such $[[p,q]]$ elements (GP= 'Grid Packing').

Assertion 11 *In the GP set the following operations are well-defined:*

$$[[p_1,q_1]] + [[p_2,q_2]] := [[p_1 + p_2, q_1 + q_2]],$$

$$\lambda[[p,q]] := [[\lambda p, \lambda q]],$$

where λ is a positive integer, and furthermore if $p_2 < p_1$ and $q_2 < q_1$ hold then

$$[[p_1,q_1]] - [[p_2,q_2]] := [[p_1 - p_2, q_1 - q_2]].$$

In the case when λ also divides p and q without a remainder term then

$$\frac{1}{\lambda}[[p,q]] := [[\frac{p}{\lambda}, \frac{q}{\lambda}]].$$

PROOF. First let us show that the statement is well defined, i.e. when we execute the above operators, the result is a GP. For one case we give the proof in detail; the other can be proved in a similar way.

Assume that

$$\frac{p_1}{q_1} \in \left(\frac{\sqrt{3}}{3}, \sqrt{3} \right) \text{ and } \frac{p_2}{q_2} \in \left(\frac{\sqrt{3}}{3}, \sqrt{3} \right).$$

We need to prove that

$$\frac{p_1 + p_2}{q_1 + q_2} \in \left(\frac{\sqrt{3}}{3}, \sqrt{3} \right).$$

According to the conditions, on the one hand

$$p_1^2 < 3q_1^2, \qquad p_2^2 < 3q_2^2,$$

and consequently $p_1^2 + p_2^2 < 3(q_1^2 + q_2^2)$ and $p_1p_2 < 3q_1q_2$, which implies

$$(p_1 + p_2)^2 < 3(q_1 + q_2)^2.$$

On the other hand,

$$q_1^2 < 3p_1^2, \qquad q_2^2 < 3p_2^2,$$

that is $q_1^2 + q_2^2 < 3(q_1^2 + q_2^2)$ and $q_1q_2 < 3p_1p_2$, which together imply that

$$(q_1 + q_2)^2 < 3(p_1 + p_2)^2.$$

□

10.2.2 Conjectural infinite grid packing sequences

The grid packing discussed in the previous section can be expressed in a recursion form.

Assertion 12 *Consider the subsequence of the approximating fraction sequence for the periodic simple continued fraction of $\frac{\sqrt{3}}{3}$, which contains every second fraction starting with the second fraction. The grid packings associated to these fractions are identical with those generated by the following recursive sequence:*

$$S_1 = [[1, 1]], \quad S_2 = [[3, 5]],$$

$$S_n = 4S_{n-1} - S_{n-2} \quad (n \geq 3).$$

PROOF. The key relationships between the continued fraction numbers and the denominators and divisors of the approximating fraction sequence in the p_k/q_k approximating sequence of $\sqrt{3}/3$ are given by

$$p_0 = 0, \quad q_0 = 1,$$

$$p_1 = 1, \quad q_1 = 1,$$

$$p_{2k} = p_{2k-1} + p_{2k-2}, \quad p_{2k+1} = 2p_{2k} + p_{2k-1},$$

$$q_{2k} = q_{2k-1} + q_{2k-2}, \quad q_{2k+1} = 2q_{2k} + q_{2k-1}.$$

Notice that the first element of the odd index subsequence in the above sequence of fractions is $1/1$, while the second element is $3/5$. In other words, if we prove that

$$p_{2k+1} = 4p_{2k-1} - p_{2k-3},$$

then the proof is complete. In fact this equality holds, since

$$p_{2k+1} = 2p_{2k} + p_{2k-1} = 2(p_{2k-1} + p_{2k-2}) + p_{2k-1} = 3p_{2k-1} + 2p_{2k-2}$$
$$= 3p_{2k-1} + (p_{2k-1} - p_{2k-3}) = 4p_{2k-1} - p_{2k-3}.$$

□

With the help of the generator function approach (which is well known in combinatorics) we can easily determine a closed expression for the values of $S_n = [[p_n, q_n]]$.

Assertion 13 *For the elements of the* $S_n = [[p_n, q_n]]$ *sequence given in Assertion 12 the following equalities hold:*

$$p_n = \frac{3-\sqrt{3}}{6}U_n + \frac{3+\sqrt{3}}{6}V_n,$$

$$q_n = \frac{\sqrt{3}-1}{2}U_n - \frac{\sqrt{3}+1}{2}V_n,$$

where $U_n = (2+\sqrt{3})^n$, $V_n = (2-\sqrt{3})^n$, *and* $n \geq 0$.

PROOF. Consider the power series

$$F(x) = \sum_{i=1}^{\infty} a_i x^i = 1 + 3x + 11x^2 + 41x^3 + \cdots$$

belonging to the $\{p_n\}_{k=1}^{\infty}$ sequence. We can easily obtain the power series

$$xF(x) = \sum_{i=1}^{\infty} a_i x^i = x + 3x^2 + 11x^3 + \cdots$$

and

$$x^2F(x) = \sum_{i=1}^{\infty} a_i x^i = x^2 + 3x^3 + \cdots,$$

that is $F(x) = 4xF(x) - x^2F(x) - x + 1$. Thus

$$F(x) = \frac{1-x}{x^2-4x+1} = \frac{1-x}{2\sqrt{3}}\left(\frac{1}{x-2-\sqrt{3}} - \frac{1}{x-2+\sqrt{3}}\right),$$

which implies the expression for p_n (using the fact that $|x| < \frac{1}{4}$). A similar line of thoughts provides the expression for q_n with the help of the

$$G(x) = \frac{x+1}{x^2 - 4x + 1}$$

generator function. □

The conjectured optimality of the above-mentioned infinite grid packing is based on the finding that the elements of this packing sequence are extremely dense packings. The densities of the related circle packings converge to $\frac{\pi}{\sqrt{12}}$. Using facts obtained on the structures of verified optimal packings and on those of the best known packings, four additional such circle packing sequences can be defined that are also conjectured to be optimal for each of their elements.

Assertion 14 *Consider the* $\{A_i\}, \{B_i\}, \{C_i\}, \{D_i\}$ *grid packing sequences, where* $(i \in \mathbb{Z}^+)$*:*

$$A_i := 2S_i, \qquad C_i := B_i/2,$$

$$B_i := S_i + S_{i+1}, \text{ and } D_i := C_i + [3,5].$$

Then

$$\lim_{i\to\infty} d_{X\,i} = \frac{\pi}{\sqrt{12}},$$

where $X \in \{S, A, B, C, D\}$ *and* $d_{X\,i}$ *denotes the density of the given circle packing.*

PROOF. Denote the number of circles in a circle packing by $n(X_i)$ and the minimal distance between the points of the corresponding point arrangement problem by $m(X_i)$.

a) Consider first the case $\lim\limits_{i\to\infty} d_{S_i} = \frac{\pi}{\sqrt{12}}$.

Due to $S_i = [s_{1i}, s_{2i}]$, $n(S_i) = \frac{(s_{1i}+1)(s_{2i}+1)}{2}$, and $m(S_i) = \frac{\sqrt{s_{1i}^2 + s_{2i}^2}}{s_{1i}s_{2i}}$, we get

$$\lim_{i\to\infty} d_{S_i} = \lim_{i\to\infty} n(S_i)\pi \frac{m(S_i)^2}{4(m(S_i)+1)^2}$$

$$= \lim_{i\to\infty} \frac{\pi}{4} \frac{\left(\frac{1}{s_{2i}}+1\right)\left(\frac{1}{s_{1i}}+1\right)}{2} \frac{\frac{s_{1i}}{s_{2i}} + \frac{s_{2i}}{s_{1i}}}{(1+m(S_i))^2} = \frac{\pi}{4}\frac{1}{2}\frac{4\sqrt{3}}{3} = \frac{\pi}{\sqrt{12}}.$$

b) The next case is $\lim\limits_{i\to\infty} d_{A_i} = \frac{\pi}{\sqrt{12}}$.

Now start with the relations $A_i = [2s_{1i}, 2s_{2i}]$, $n(A_i) = 2s_{1i}s_{2i} + s_{1i} + s_{2i} + 1$, and

$$m(A_i) = \frac{\sqrt{s_{1i}^2 + s_{2i}^2}}{2s_{1i}s_{2i}}.$$

These imply that

$$\lim_{i\to\infty} d_{A_i} = \lim_{i\to\infty} (2s_{1_i}s_{2_i} + s_{1_i} + s_{2_i} + 1)\pi \frac{s_{1_i}^2 + s_{2_i}^2}{4\left(\sqrt{s_{1_i}^2 + s_{2_i}^2} + 2s_{1_i}s_{2_i}\right)^2} =$$

$$\lim_{i\to\infty} \left(2 + \frac{1}{s_{2_i}} + \frac{1}{s_{1_i}} + \frac{1}{s_{1_i}s_{2_i}}\right)\pi \frac{\frac{s_{1_i}}{s_{2_i}} + \frac{s_{2_i}}{s_{1_i}}}{4\left(2 + m(S_i)\right)^2} = 2\pi \frac{\frac{\sqrt{3}}{3} + \sqrt{3}}{4(2)^2} = \frac{\pi}{\sqrt{12}}.$$

c) The third case is $\lim_{i\to\infty} d_{B_i} = \frac{\pi}{\sqrt{12}}$.

Since $B_i = [b_{1_i}, b_{2_i}]$, $n(B_i) = \frac{(b_{1_i}+1)(b_{2_i}+1)+1}{2}$, and $m(B_i) = \sqrt{\frac{1}{b_{1_i}^2} + \frac{1}{b_{2_i}^2}}$, we can evaluate

$$\lim_{i\to\infty} d_{B_i} = \lim_{i\to\infty} \frac{\pi}{4} \frac{(b_{1_i}+1)(b_{2_i}+1)+1}{2} \frac{b_{1_i}^2 + b_{2_i}^2}{\left(\sqrt{b_{1_i}^2 + b_{2_i}^2} + b_{1_i}b_{2_i}\right)^2} =$$

$$\lim_{i\to\infty} \frac{\pi}{4} \frac{\left(1 + \frac{1}{b_{1_i}}\right)\left(1 + \frac{1}{b_{2_i}}\right) + \frac{1}{b_{1_i}b_{2_i}}}{2} \frac{\frac{b_{1_i}}{b_{2_i}} + \frac{b_{2_i}}{b_{1_i}}}{\left(1 + m(B_i)\right)^2} = \frac{\pi}{4} \frac{1}{2} \frac{\frac{1}{\sqrt{3}} + \sqrt{3}}{1} = \frac{\pi}{\sqrt{12}}.$$

d) Next consider $\lim_{i\to\infty} d_{C_i} = \frac{\pi}{\sqrt{12}}$. According to the definitions, $C_i = \frac{1}{2}\left[b_{1_i}, b_{2_i}\right]$,

$$n(C_i) = \frac{\left(\frac{b_{1_i}}{2} + 1\right)\left(\frac{b_{2_i}}{2} + 1\right)}{2},$$

and $m(C_i) = 2m(B_i)$. Now the calculation can be made as

$$\lim_{i\to\infty} d_{C_i} = \frac{\pi}{4} \frac{(b_{1_i}+2)(b_{2_i}+2)}{2} \frac{b_{1_i}^2 + b_{2_i}^2}{\left(2\sqrt{b_{1_i}^2 + b_{2_i}^2} + b_{1_i}b_{2_i}\right)^2} =$$

$$\lim_{i\to\infty} \frac{\pi}{4} \frac{\left(1 + \frac{2}{b_{1_i}}\right)\left(1 + \frac{2}{b_{2_i}}\right)}{2} \frac{\frac{b_{1_i}}{b_{2_i}} + \frac{b_{2_i}}{b_{1_i}}}{(1 + m(C_i))^2} = \frac{\pi}{8} \frac{\left(\frac{1}{\sqrt{3}} + \sqrt{3}\right)}{1} = \frac{\pi}{\sqrt{12}}.$$

e) The last case studied yields the asymptotic value $\lim_{i\to\infty} d_{D_i} = \frac{\pi}{\sqrt{12}}$.

Similar to the earlier procedures, use the definitions $D_i = [c_{1_i} + 3, c_{2_i} + 5]$,

$$n(D_i) = \frac{(c_{1_i}+4)(c_{2_i}+6)}{2},$$

and $m(D_i) = \sqrt{\frac{1}{(c_{1_i}+3)^2} + \frac{1}{(c_{2_i}+5)^2}}$. Apply them in the calculation of $\lim_{i\to\infty} d_{D_i} =$

Table 10.1. Optimal grid packings.

p	q	n	Sequences
1	1	2	S_1
2	2	5	A_1
2	3	6	C_1
3	5	12	S_2
4	6	18	B_1
5	8	27	D_1

$$\lim_{i\to\infty} \frac{\pi}{4} \frac{(c_{1i}+4)(c_{2i}+6)}{2} \frac{(c_{1i}+3)^2+(c_{2i}+5)^2}{\left(\sqrt{(c_{1i}+3)^2+(c_{2i}+5)^2}+(c_{1i}+3)(c_{2i}+5)\right)^2}$$

$$= \lim_{i\to\infty} \frac{\pi}{4} \frac{\frac{(c_{1i}+4)(c_{2i}+6)}{(c_{1i}+3)(c_{2i}+5)}}{2} \frac{\frac{c_{1i}+3}{c_{2i}+5}+\frac{c_{2i}+5}{c_{1i}+3}}{(1+m(D_i))^2} = \frac{\pi}{4}\frac{1}{2}\frac{\frac{1}{\sqrt{3}}+\sqrt{3}}{1} = \frac{\pi}{\sqrt{12}}.$$

□

Table 10.1 contains data on such grid packings that are verified to be optimal. We have also given the names of the related grid packing sequences of Assertion 14. Table 10.1 shows also that the $[[1,1]]$ and $[[2,2]]$ grid packings are optimal. The next assertion tells us that these are the only optimal grid packings of the sequence $[[p,p]]$.

Assertion 15 *If* $p > 2$, *then* $[[p,p]]$ *cannot be optimal.*

PROOF. Let us denote the number of the points of $[[p,p]]$ by $n([[p,p]])$, and the minimal distance between the points by $m([[p,p]])$. Then

$$n([[p,p]]) = \left\lceil \frac{(p+1)^2}{2} \right\rceil, \qquad m([[p,p]]) = \frac{\sqrt{2}}{p}.$$

a) If p is an even number, then $p = \sqrt{2n-1}-1$. If $p \geq 14$ (or $n([[p,p]]) \geq 112$, then after a short calculation using Theorem 3.1, we obtained the result

$$m([[p,p]]) = \frac{\sqrt{2}}{\sqrt{2n-1}-1} < \sqrt{\frac{2}{\sqrt{3}n}} < m_n.$$

If $p = 4, 6, 8, 10,$ or 12, then from the Table 10.2 you can see that $[[p,p]]$ cannot be optimal because in the cases when $p = 4, 6$, $m([[p,p]]) < \overline{m}_n$ and for $p = 8, 10, 12$, $m([[p,p]]) < m_n$.

b) If p is an odd number, then $p = \sqrt{2n}-1$. In this case if $p \geq 15$ (or $n \geq 104$), then using Theorem 3.1,

Table 10.2. Details for the grid packing $[[p,p]]$ when $2 < p < 14$ is an even number.

p	$n([[p,p]])$	$m([[p,p]])$	m_n
4	13	0.35355	0.36609
6	25	0.23570	0.25000
8	41	0.17677	0.18609
10	61	0.14142	0.14854
12	85	0.11785	0.12511

Table 10.3. Details for the grid packing $[[p,p]]$ when $2 < p < 14$ is an odd number.

p	$n([[p,p]])$	$m([[p,p]])$	m_n
3	8	0.47140	0.51763
5	18	0.28284	0.30046
7	32	0.20203	0.21313
11	72	0.12856	0.13569
13	98	0.10878	0.11610

$$m([[p,p]]) = \frac{\sqrt{2}}{\sqrt{2n}-1} < \sqrt{\frac{2}{\sqrt{3}n}} < m_n.$$

If $p = 3, 5, 7, 11$, and 13, then based on Table 10.3 we may state that $[[p,p]]$ cannot be optimal because for the cases $p = 3, 5$ $m([[p,p]]) < \overline{m}_n$ and for the best known values for $p = 7, 11$, and 13 $m([[p,p]]) < m_n$. □

10.3 Improved theoretical lower bounds based on pattern classes

The pattern classes discussed earlier are also useful for improving the theoretical lower bounds.

Theorem 10.1. *The m_n value is greater than or equal to*

$$\max\left(L_1(n), L_2(n), L_{3a}(n), L_{3b}(n), L_4(n), L_5(n), L_6(n), L_7(n), L_8(n), L_9(n)\right),$$

where

$$L_1(n) = \frac{1}{\lceil\sqrt{n}\rceil - 1},$$

$$L_2(n) = \frac{1}{\lceil\sqrt{n+1}\rceil - 3 + \sqrt{2+\sqrt{3}}},$$

$$L_{3a}(n) = \frac{1}{\lceil\sqrt{n+2}\rceil - 2 + \frac{1}{2}\sqrt{3}},$$

$$L_{3b}(n) = \frac{1}{\lceil\sqrt{n+2}\rceil - 5 + 2\sqrt{2+\sqrt{3}}},$$

$$L_4(n) = \frac{1}{\lceil\sqrt{n+3}\rceil - 3 + \sqrt{3}},$$

$$L_5(n) = \frac{1}{\lceil\sqrt{n+4}\rceil - 3 + \sqrt{3}},$$

$$L_6(n) = \frac{1}{\lceil\sqrt{n+5}\rceil - 4 + 3\sqrt{3}/2},$$

$$L_7(n) = \frac{k^2 - k - \sqrt{2k}}{k^3 - 2k^2}, \text{ if } n = k(k+1), \text{ otherwise } 0,$$

$$L_8(n) = \max_i \left\{ \sqrt{\frac{1}{p_i^2} + \frac{1}{q_i^2}} \right\}, \text{ if } n = \left\lceil \frac{(p_i+1)(q_i+1)}{2} \right\rceil, \; p_i^2 \le 3q_i^2 \text{ and } q_i^2 \le 3p_i^2, i \in \mathbb{N}, \text{ otherwise } 0,$$

$$L_9(n) = \sqrt{\frac{2}{\sqrt{3}n}}.$$

PROOF. The proof is constructive. Let us consider the maximum of the lower bounds based on the previous determined pattern classes, structure classes, grid packings, and on the lower bound of Theorem 3.1.

In order to calculate k in the function of n from the $n = f(k)$ expression, after replacing them in l_k, we get those functions that are in the theorem. In the definition of L_8 we used the maximum values, because there can be more than one description of a positive number in the form of $n = \left\lceil \frac{(p_i+1)(q_i+1)}{2} \right\rceil, p_i^2 \le 3q_i^2$ and $q_i^2 \le 3p_i^2$. □

We may examine the previous lower bound and the upper bound using the statement of Theorem 3.2 on the upper bound of $\overline{r}_n$ in a diagram in Figure 10.12. The graph of the function was drawn based on discrete points, but we used a fraction-linear function to display it in a nicer form.

10.3.1 Sharp, constant lower and upper bounds for the density of circle packing problems

The lower and upper bounds we obtained implied sharp bounds for the density of circle packings (independent of the number of the circles).

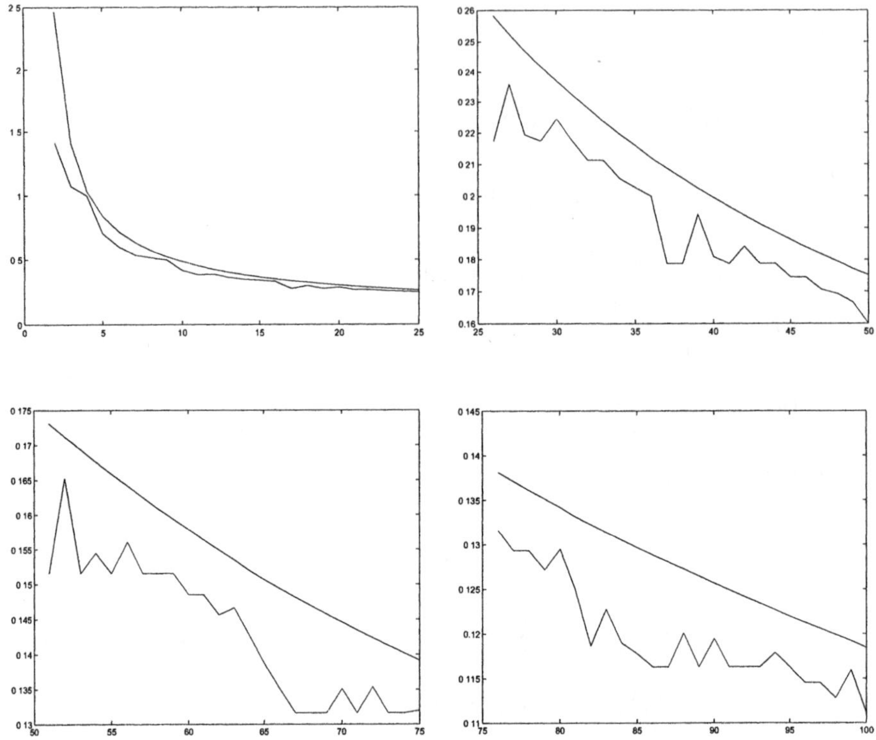

Fig. 10.12. Lower and upper bounds of $\overline{m}_n$ for $n = 2 - 100$ circles.

Theorem 10.2. *For every* $n \geq 2$ *integer number*

$$(3 - 2\sqrt{2})\pi \leq \ d_n(\overline{r}_n, [0,1]^2) < \frac{\pi}{\sqrt{12}}$$

hold, and the bounds are sharp.

PROOF. In Theorem 3.1 we proved that $\sqrt{\frac{2}{\sqrt{3}n}} < \overline{m}_n$. This lower bound implies the following lower bound for the density:

$$\frac{n\pi}{(2 + \sqrt[4]{12n^2})^2} < \overline{d}_n(\overline{r}_n, [0,1]^2).$$

As the densities of the optimal packings are known for up to $n = 30$, it easy to check that for $n = 13$ circles the density of the optimal packing is greater than or equal to

$$\overline{d}_n(\overline{r}_2, [0,1]^2) = (3 - 2\sqrt{2})\pi \approx 0.539$$

(see Table 10.4). With $n > 13$, after a short calculation the following inequality can be proved:

Table 10.4. The density for packings up to $n = 13$ circles.

n	$\approx d_n(\overline{r}_n, [0,1]^2)$	n	$\approx d_n(\overline{r}_n, [0,1]^2)$
2	0.5390120845	8	0.7309638253
3	0.6096448087	9	0.7853981634
4	0.7853981634	10	0.6900357853
5	0.6737651056	11	0.7007415778
6	0.6639569095	12	0.7384682239
7	0.6693108268	13	0.7332646949

$$(3 - 2\sqrt{2})\pi < \frac{n\pi}{(2 + \sqrt[4]{12n^2})^2} < \overline{d}_n(\overline{r}_n, [0,1]^2),$$

and the lower bound is sharp because $\overline{d}_n(\overline{r}_2, [0,1]^2) = (3 - 2\sqrt{2})\pi$.

The upper bound for the density is equivalent to the following inequality:

$$\overline{m}_n < f_1(n) = \frac{2 + \sqrt{2\sqrt{3}n}}{\sqrt{3}n - 2}.$$

Based on Theorem 3.2 we know that

$$\overline{m}_n \le f_2(n) = \frac{1 + \sqrt{1 + (n-1)\frac{2}{\sqrt{3}}}}{n - 1}.$$

With a short calculation it can be proved that $f_2(n) < f_1(n)$, for $n \ge 2$.

The sharpness of the bounds means that we cannot improve them, so if we have a constant smaller than $\pi/\sqrt{12}$, then we can determine a circle packing for which the density is greater than the given value. Hence it is enough to find a circle packing series, where the density of the circle packings tends to $\pi/\sqrt{12}$. We have already examined this kind of series (see Subsection 10.2.2), which yields the required value. □

11

Minimal Polynomials of Point Arrangements

When investigating the existence of circle packings with a given structure we saw that for the circle packings—or point arrangements—we could represent them by a corresponding quadratical system of equations. If this system of equations has an algebraic solution in $[0,1]^{2n} \times (0, \sqrt{2}]$ then the Groebner basis of the polynomials will be upper triangular in its variables. If we order the variables of polynomials in such a way that the variable m comes first, then the Groebner basis will contain a polynomial in m, the smallest positive root of which is the minimal distance of the point arrangement. A polynomial with this property and of minimal degree is the minimal polynomial of the packing [93, 94].

11.1 Determining the minimal polynomial for a point arrangement

Sometimes it is an easy task to determine the minimal polynomial of a point arrangement. For example, let us look at the following cases:

a) the minimal polynomials for the point arrangements in the pattern class **PAT**(k^2) are obviously

$$(k-1)m-1,$$

b) the minimal polynomials for the point arrangements in the pattern class **PAT**$_1(k^2-2)$ are

$$(4k^2-16k+13)m^2-(8k-16)m+4,$$

c) and finally the minimal polynomials for the point arrangements in the pattern class **PAT**$_2(k(k+1))$ are

$$k^3(k-2)m^2-2k^2(k-1)m+k^2+1.$$

The degree of difficulty of determining the minimal polynomials for a circle packing is related to the structure of circle packing.

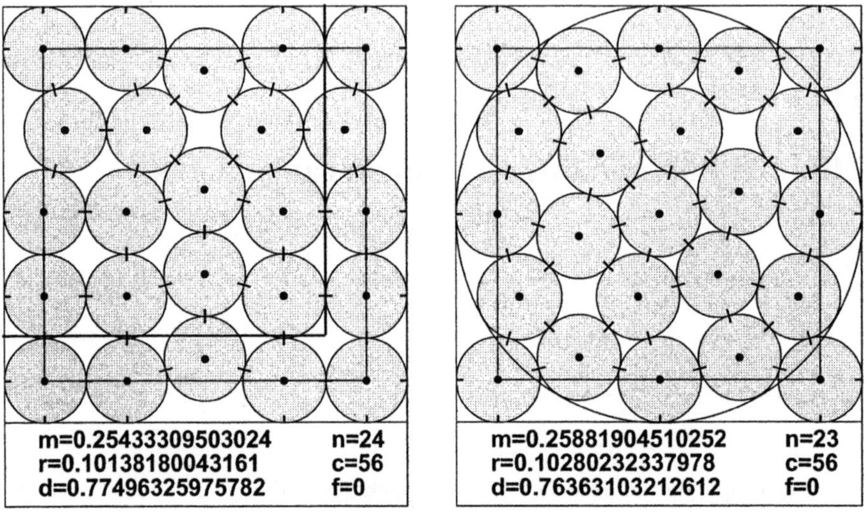

Fig. 11.1. Two examples of optimal substructures in optimal circle packings.

First we are going to investigate a nice example from the viewpoint of determining minimal polynomials, when the actual circle packing contains optimal substructures. In such situations the minimal polynomials can be found from the generalized minimal polynomials of substructures using resultant theory [116].

11.1.1 Optimal substructures

First of all, let us examine the concepts of optimal substructures and generalized minimal polynomials.

Definition 11.1. *We call a circle packing (or the related point arrangement) an optimal substructure in the* $X \subset [0,1]$ *compact set, if the density* $d_{n'}(r_{n'}, [0,1]^2)$ *is maximal in* X*, where* n' *denotes the number of circles in* X.

In Figure 11.1 there is a delightful example where X is a square with sides parallel to the original square in the upper left corner (the former is called a fitting square). We can also see an example when the set X is a circle. In the upper left-hand corner of the square with 24 packed circles there is an optimal substructure with 15 circles. When we consider the optimal packing of 23 circles in the square, then the circle inscribed with a radius of $1/2$ contains an optimal substructure, since it is well known that it is the densest configuration of 19 circles in a circle [26].

In is interesting to note that optimal substructures are not so rare in optimal circle packings. The containment graph in Figure 11.2 shows, for the circle packings with

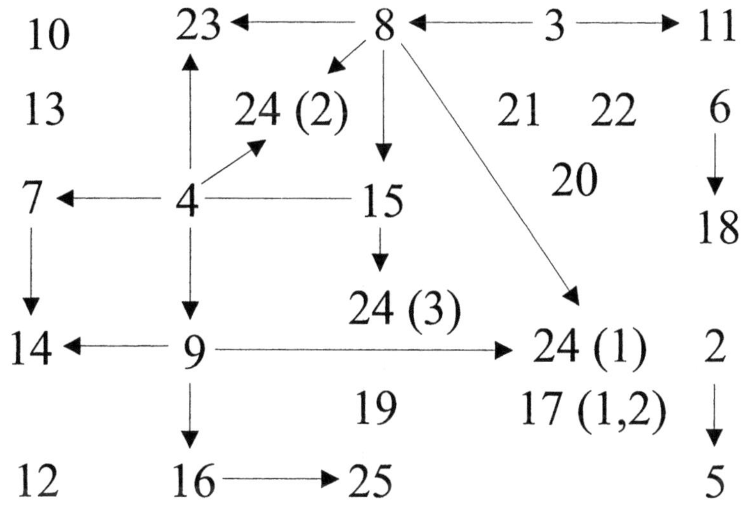

Fig. 11.2. The containment graph for the optimal circle packings with $n = 2 - 27$.

$n = 2 - 27$, what kind of optimal substructures exist. The edges of the graph are numbers which tell us the number of circles in a circle packing. When an arrow points from a to b, then the meaning is that the optimal circle packing with b circles contains an optimal circle packing with a circles, where X is a square. The numbers in brackets indicate the numbers of different optimal solutions.

Definition 11.2. *The $p_n^I(x)$ polynomial is called a generalized minimal polynomial if $x \in \{r, m, s, \sigma\}$ and $I \in \{S, \Sigma, R, M\}$, $\overline{x}_n$ is the smallest positive root of the polynomial $p_n^I(x)$, and the degree of the polynomial is minimal. Let $P_n(x) = p_n^1(x)$.*

Assertion 16 *The relationships between the generalized polynomials can be described by the formulas contained in Table 11.1.*

PROOF. The given relations can be readily found by using the relations stated in Corollary 2.15. □

To see how the formulas of Assertion 16 work, let us go through a brief example.

Example 11.3. Consider the problem of finding the $p_{11}^S(r)$ generalized minimal polynomial based on the known minimal polynomial

$$P_{11}(m) = m^8 + 8m^7 - 22m^6 + 20m^5 + 18m^4 - 24m^3 - 24m^2 + 32m - 8.$$

It is easy to check that

$$p_n^{\Sigma}(m) = P_n(m/\Sigma)\Sigma^{\deg P_n},$$

$\mathbf{p_n^S}(r) = p_n^{\Sigma:=S-2r}(m := 2r)$	$\mathbf{p_n^\Sigma}(m) = p_n^{R:=\Sigma+m}(s := \frac{m}{2})$
$p_n^{R:=S}(s := r)$	$p_n^{M:=\Sigma}(\sigma := m)$
$p_n^{M:=S-2r}(\sigma := 2r)$	$p_n^{S:=\Sigma+m}(r := \frac{m}{2})$
$\mathbf{p_n^R}(s) = p_n^{M:=R-2s}(\sigma := 2r)$	$\mathbf{p_n^M}(\sigma) = p_n^{S:=M+\sigma}(r := \frac{\sigma}{2})$
$p_n^{S:=R}(r := s)$	$p_n^{\Sigma:=M}(m := \sigma)$
$p_n^{\Sigma:=R-2s}(m := 2s)$	$p_n^{R:=M+m}(s := \frac{\sigma}{2})$

Table 11.1. Relationships between the generalized minimal polynomials.

hence

$$p_{11}^\Sigma(m) = m^8 + 8m^7\Sigma - 22m^6\Sigma^2 + 20m^5\Sigma^3 + 18m^4\Sigma^4 - 24m^3\Sigma^5 - 24m^2\Sigma^6$$
$$+32m\Sigma^7 - 8\Sigma^8.$$

Using the fact that $p_n^S(r) = p_n^{\Sigma:=S-2r}(m := 2r)$ we get

$$p_{11}^S(r) = p_{11}^{\Sigma:=S-2r}(m := 2r) =$$

$$(2r)^8+8(2r)^7(S-2r)-22(2r)^6(S-2r)^2+20(2r)^5(S-2r)^3+18(2r)^4(S-2r)^4$$
$$-24(2r)^3(S-2r)^5 - 24(2r)^2(S-2r)^6 + 32(2r)(S-2r)^7 - 8(S-2r)^8 =$$
$$-18176r^8 + 45056r^7S - 63360r^6S^2 + 56192r^5S^3 - 30432r^4S^4 + 9920r^3S^5$$
$$-1888r^2S^6 + 192rS^7 - 8S^8.$$

The minimal polynomial we desire can be obtained by dividing the above polynomial by 8: $p_{11}^S(r) =$

$$2272r^8 - 5632r^7S + 7920r^6S^2 - 7024r^5S^3 + 3804r^4S^4 - 1240r^3S^5 + 236r^2S^6$$
$$-24rS^7 + S^8.$$

Theorem 11.4. *Consider a point arrangement problem in the unit square. Assume that the point arrangement has $N \geq 2$ optimal substructures in the fitting squares of side length $\Sigma_1, \Sigma_2, \ldots,$ and Σ_N, respectively. If f is such a polynomial that has such $1 \leq i, j \leq N$ indices for which $\Sigma_i = f(\Sigma_j)$, then the $p_n^\Sigma(m)$ minimal polynomial can be calculated from the minimal polynomials of the i-th and j-th optimal substructures that is*

$$p_n^\Sigma(m) = \mathrm{Res}(p_{n_1}^{\Sigma_j}(m), p_{n_2}^{f(\Sigma_j)}(m), \Sigma_j)$$
$$= \det(\mathrm{Syl}(p_{n_1}^{\Sigma_j}(m), p_{n_2}^{f(\Sigma_j)}(m), \Sigma_j)).$$

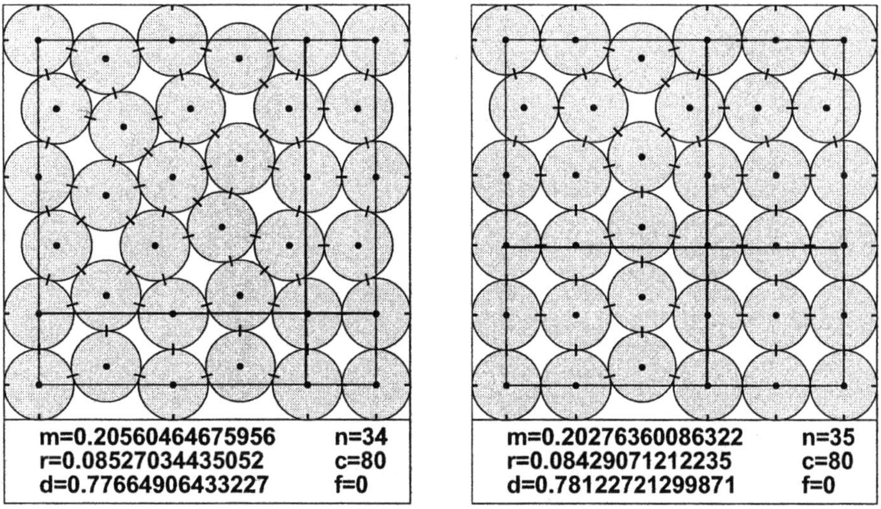

Fig. 11.3. Circle packings for $n = 34$ and $n = 35$ circles with optimal substructures indicated.

PROOF. The proof is a direct consequence of the definition of the resultant. If we can determine the relationship between the sides of the two squares containing the optimal substructures by the polynomial f, then the minimal polynomial for the circle packing defined on the square $[0,1]^2$ can be obtained from evaluating the minimal polynomials of the optimal substructures (which have the two common roots of m and Σ_j), that is by forming their resultant and eliminating Σ_j. $\square$

Example 11.5. Find the minimal polynomial $P_{34}(m)$ using the minimal polynomials of the substructures, $p_{23}^{\Sigma_1}(m)$ and $p_4^{\Sigma_2}(m)$. See Figure 11.3. In this example $f(x) = 1 - x$, hence

$$p_{23}^{\Sigma_1}(m) = 16m^4 - 16m^2\Sigma_1^2 + \Sigma_1^4, p_4^{\Sigma_2}(m) = m - \Sigma_2 = m - 1 + \Sigma_1,$$

$$P_{34}(m) = \mathrm{Res}(p_{23}^{\Sigma_1}(m), p_4^{1-\Sigma_1}(m), \Sigma_1)$$

$$= \begin{vmatrix} 1 & 0 & 0 & 0 & 1 \\ m-1 & 1 & 0 & 0 & 0 \\ 0 & m-1 & 1 & 0 & -16m^2 \\ 0 & 0 & m-1 & 1 & 0 \\ 0 & 0 & 0 & m-1 & 16m^4 \end{vmatrix} = m^4 + 28m^3 - 10m^2 - 4m + 1.$$

The relationship given in the next assertion can also be used to find minimal polynomials.

Assertion 17 *Consider the $P_n(m)$ minimal polynomial, and assume that*

$$m_n = \frac{am_{n'} + b}{cm_{n'} + d} \quad \text{that is,} \quad m_{n'} = \frac{b - dm_n}{cm_n - a},$$

where a, b, c, and d are real numbers. Then the $P_{n'}(m)$ minimal polynomial can be determined in the following way:

$$P_{n'}(m) = P_n\left(\frac{am+b}{cm+d}\right)(cm+d)^{\deg P_n}.$$

PROOF. It is easy to see that $P_n\left(\frac{am+b}{cm+d}\right)(cm+d)^{\deg P_n}$ is in fact a polynomial, and $m_{n'}$ is a root of it. Furthermore it is a minimal polynomial at the same time, since otherwise there would exist another R polynomial, for which $R(m_{n'}) = 0$ and

$$\deg R < \deg P_n\left(\frac{am+b}{cm+d}\right)(cm+d)^{\deg P_n}.$$

However, it is impossible, since

$$(\deg R =) \deg R\left(\frac{b-dm}{cm-a}\right)(cm-a)^{\deg R} < \deg P_n$$

would then hold. This contradicts the fact that $P_n(m)$ is a minimal polynomial. □

Example 11.6. Find the $P_{35}(m)$ minimal polynomial (see the related circle packing diagram in Figure 11.3.

We will determine the $P_{35}(m)$ minimal polynomial in two different ways, first from the application of Theorem 11.4, and secondly using Assertion 17.

a) Applying Theorem 11.4, and making use of the $p_{15}^{\Sigma_1}(m)$ and $p_9^{\Sigma_2}(m)$ minimal polynomials we get

$$f(x) = 1 - x,$$

$$p_{15}^{\Sigma_1}(m) = 2m^4 - 4m^3\Sigma_1 - 2m^2\Sigma_1^2 + 4m\Sigma_1^3 - \Sigma_1^4,$$

$$p_9^{\Sigma_2}(m) = 2m - \Sigma_2 = 2m - 1 + \Sigma_1,$$

$$P_{35}(m) = \mathrm{Res}(p_{15}^{\Sigma_1}(m), p_9^{1-\Sigma_1}(m), \Sigma_1) =$$

$$\begin{vmatrix} 1 & 0 & 0 & 0 & -1 \\ 2m-1 & 1 & 0 & 0 & 4m \\ 0 & 2m-1 & 1 & 0 & -2m^2 \\ 0 & 0 & 2m-1 & 1 & -4m^3 \\ 0 & 0 & 0 & 2m-1 & 2m^4 \end{vmatrix} = 46m^4 - 84m^3 + 50m^2 - 12m + 1.$$

Of course, the same result can be obtained if we use Assertion 17.

b) Since

$$P_{24}(m) = m^4 - 16m^3 + 20m^2 - 8m + 1$$

and

$$m_{35} = 2r_{24},$$

then

$$m_{35} = 2r_{24} = \frac{m_{24}}{m_{24}+1}, \text{ and consequently } m_{24} = \frac{m_{35}}{1-m_{35}},$$

which implies that

$$P_{35}(m) = P_{24}\left(\frac{m}{1-m}\right)(1-m)^4 = 46m^4 - 84m^3 + 50m^2 - 12m + 1.$$

11.1.2 Examining the general case

In the general case, when the circle packing does not contain any special features like, for example, a symmetry or the existence of optimal substructures, then one method for determining the minimal polynomial is to choose that polynomial from the Gröbner basis [13] of the polynomial system, which depends only on m. This should be ordered in such a way that m is the smallest. Then the determination of the full Gröbner basis is not necessary (and is in fact needless), since some symbolic algebra systems (such as Maple or Mathematica) support the determination of the minimal polynomial from the respective Gröbner basis. In certain cases this method is applicable for the determination of the minimal polynomials of circle packings.

First we shall give a short overview of the most important concepts of Gröbner basis theory.

Definition 11.7. x^α *is a* monomial *of the variables* $x_1, x_2, \ldots, x_n$, *if it can be written in the form* $x^\alpha = x_1^{\alpha_1} x_2^{\alpha_2} \cdots x_n^{\alpha_n}$, *where* $\alpha_1, \alpha_2, \ldots, \alpha_n \in \mathbb{N}$ *and* $\alpha = (\alpha_1, \ldots, \alpha_n)$. *The* monomial order *is defined by* $|\alpha| = \alpha_1 + \alpha_2 + \cdots + \alpha_n$. *Should* $\alpha = (0, \ldots, 0)$, *then* $x^\alpha = 1$.

Definition 11.8. *A polynomial defined over a* $\mathbb{K}$ *field is a finite linear combination of the* x^α *monomials with coefficients from* $\mathbb{K}$. *The* degree of a polynomial *is the degree of the maximal degree monomial that has a non-zero coefficient.*

Definition 11.9. *A polynomial ring over* $\mathbb{K}$ *(denoted by* $\mathbb{K}[x_1, x_2, ..., x_n]$*) is defined as the set of all* $x_1, x_2, ..., x_n$ *variable polynomials over* $\mathbb{K}$ *with the operations of addition and multiplication applied in the obvious way.*

The concept of a ring in Definition 11.9 is correct, since it is easy to prove that $\mathbb{K}[x_1, x_2, ..., x_n]$ is a commutative ring with a unit element.

Definition 11.10. *We call a non-empty subset I of a $\mathbb{K}[x_1, x_2, ..., x_n]$ commutative ring an* ideal, *if* $\forall a, b \in I \Rightarrow a + (-b) \in I$ *and* $\forall r \in \mathbb{K}[x_1, x_2, ..., x_n], a \in I \Rightarrow r \cdot a \in I, a \cdot r \in I$.

Definition 11.11. *Let $P = \{p_1, p_2, \ldots, p_k\}$ be a multidimensional system of polynomials. Then the set*

$$(P) = \left\{ \sum_{i=1}^{k} a_i p_i : a_i \in \mathbb{K}[x_1, x_2, \ldots, x_n] \right\}$$

will be obviously an ideal of it. It is called an ideal generated by P. *The set P is then the* basis of the ideal, *and hence the* ideal is finitely generated.

According to a basic theorem by *Hilbert* all ideals of the $\mathbb{K}[x_1, x_2, ..., x_n]$ polynomial ring are finitely generated, and can be constructed in the way given in Definition 11.11.

Notice that there exists a bijective mapping between the x^α monomials and the $\alpha = (\alpha_1, \ldots, \alpha_n) \in \mathbb{N}^n$ ordered n-tuples. This means that when we can define a $\preceq$ order relation on the elements of $\mathbb{N}^n$ it induces an order on the monomials too:

$$\alpha \preceq \beta \Leftrightarrow x^\alpha \preceq x^\beta.$$

Definition 11.12. *The order $\preceq$ defined on the monomials is called* feasible *if the equivalent order on $\mathbb{N}^n$ satisfies the following conditions:*

a, $\preceq$ *is a linear order,*
b, $\alpha \preceq \beta$ *and* $\gamma \in \mathbb{N}^n$ *then* $\alpha + \gamma \preceq \beta + \gamma$, *and*
c, $\preceq$ *is a total order in* $\mathbb{N}^n$.

The *leading term* of a polynomial respecting an order $\preceq$ is its monomial whose degree is the largest degree coefficient of the polynomial.

Several orders can be defined on monomials, one of them being the $\preceq_L$ lexicographic order:

$$x^\alpha = x_1^{\alpha_1} x_2^{\alpha_2} \cdots x_n^{\alpha_n} \preceq x_1^{\beta_1} x_2^{\beta_2} \cdots x_n^{\beta_n} = x^\beta$$

if and only if $(\alpha_1, \alpha_2, ..., \alpha_n) = (\beta_1, \beta_2, ..., \beta_n)$, or there exists such a positive integer i that $\alpha_j = \beta_j$ $(1 \leq j \leq i - 1)$ and $\alpha_i < \beta_i$.

Definition 11.13. *Let $\preceq$ be a feasible order on the space of the monomials of the variables $x_1, x_2, \ldots, x_n$. A G non-empty subset of an I ideal of the $\mathbb{K}[x_1, x_2, \ldots, x_n]$ polynomial ring is called the* Gröbner basis *of the I ideal for the $\preceq$ order if each polynomial of the ideal has such a leading term (respecting the order) that can be divided by the leading term of a polynomial for G.*

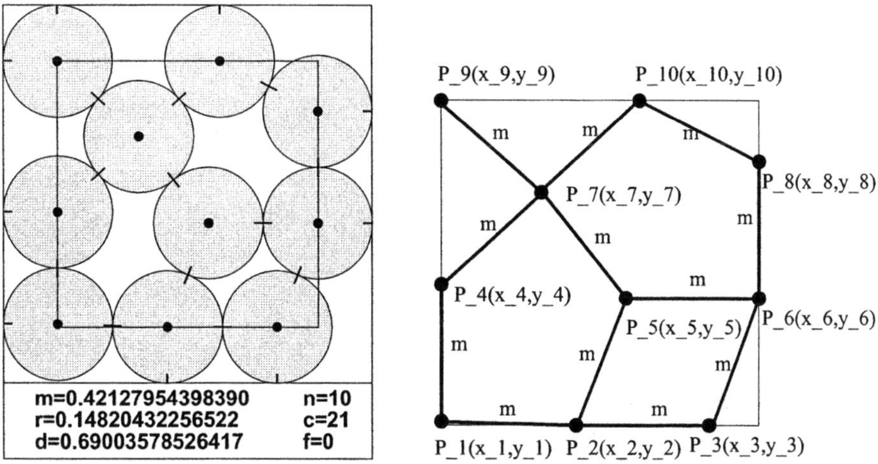

Fig. 11.4. The optimal packing for 10 circles in the unit square.

The theory of Gröbner bases is important from the point of view of the solvability of systems of polynomials. Since an ideal I generated by a P system of polynomials and its Gröbner basis is generated by the ideal I, the $P = 0$ and $G = 0$ polynomial systems are solvable at the same time (or neither of them) and, furthermore, the solution sets are identical as well.

B. Buchberger first introduced and defined the notion of a Gröbner basis, named in 1965 in honour of his PhD supervisor, *W. Gröbner* (1899–1980) . A similar notion was suggested by *H. Hironaka* in 1964, still Buchberger also provided an algorithm to determine the Gröbner basis. This algorithm is a generalization of both the Euclidian algorithm and the Gauss elimination approach.

In the investigation of the existence of a circle packing structure the theorem of the theory has an important role because it states that a P system of polynomials is solvable if and only if 1 is not contained in the Gröbner basis of the ideal generated by P. In addition to this we can prove the nonexistence of a circle packing structure when we prove that all the polynomials in the Gröbner basis of the polynomial system associated with the structure has roots (working with complex coefficients) that they are not in the $[0, 1]$ interval for a variable.

Next we will show how Gröbner basis theory can be used to solve a nonexistence problem with the help of the Maple symbolic algebra system.

Example 11.14. Proving the nonexistence of the structure of a circle packing.

The optimal solution for the 10 circle packing problem elaborated on in the Introduction can be seen in Figure 11.4. An interesting point about the packing structure is that the lower left circle seems to be in the corner, but in fact it touches only one of the sides. By using Gröbner basis theory we can prove that a circle packing structure that differs from the one in Figure 11.4 only in that the lower left circle is exactly in the corner does not exist.

Although G. Valette published a structure [126] where point P_1 is in the lower left-hand corner, in his case the distance of the P_3 and P_6 points is larger than m. In a monograph on the field, in the Handbook of Discrete and Computational Geometry [34] the figure depicting the optimal packing of 10 circles is erroneous, since the above-mentioned circle exactly fits the two sides in the corner.

The structure of the modified "circle packing" can be given by the following system of equations using the notation of Figure 11.4:

$$\begin{array}{rr}
(x_1-x_2)^2+(y_1-y_2)^2=m^2 & (x_1-x_4)^2+(y_1-y_4)^2=m^2\\
(x_2-x_3)^2+(y_2-y_3)^2=m^2 & (x_2-x_5)^2+(y_2-y_5)^2=m^2\\
(x_5-x_6)^2+(y_5-y_6)^2=m^2 & (x_3-x_6)^2+(y_3-y_6)^2=m^2\\
(x_4-x_7)^2+(y_4-y_7)^2=m^2 & (x_5-x_7)^2+(y_5-y_7)^2=m^2\\
(x_7-x_9)^2+(y_7-y_9)^2=m^2 & (x_7-x_{10})^2+(y_7-y_{10})^2=m^2\\
(x_8-x_{10})^2+(y_8-y_{10})^2=m^2 & (x_6-x_8)^2+(y_6-y_8)^2=m^2\\
x_1=x_4=x_9=y_2=y_3=\mathbf{y_1}=0 & x_6=x_8=y_9=y_{10}=1.
\end{array}$$

An important change is made in the highlighted $\mathbf{y_1}$ variable by setting its value to zero. This modification means that the respective circle is exactly in the lower left-hand corner of the unit square.

Let us use some geometric properties of circle packings, with which we can reduce the system of equations. It is easy to see that $x_2=m$, $y_4=y_1+m$, $x_3=x_2+m$ and $y_8=y_6+m$. The quadrangle $P_2P_3P_5P_6$ is in fact a rhombus, and hence $x_5=1-m$ and $y_5=y_6$. The triangles $P_4P_7P_9$ and $P_9P_7P_{10}$ are isosceles, thus the P_4, P_7, and P_{10} points are collinear, meaning that $y_7=(1+y_1+m)/2$ and $x_7=x_{10}/2$.

Using the earlier observations, we can eliminate all variables with the exception of x_2, x_{10}, y_5 and m. The system of equations reduces then to:

$$\begin{aligned}
(1-2m)^2+y_5^2&=m^2,\\
x_{10}^2+(1-m)^2&=(2m)^2,\\
(1-x_{10})^2+(1-y_5-m)^2&=m^2,\\
(2-2m-x_{10})^2+(2y_5-1-m)^2&=(2m)^2.
\end{aligned}$$

The Gröbner basis of the polynomial system $P=\{(1-2m)^2+y_5^2-m^2, x_{10}^2+(1-m)^2-(2m)^2, (1-x_{10})^2+(1-y_5-m)^2-m^2, (2-2m-x_{10})^2+(2y_5-1-m)^2-(2m)^2\}$ is $G=\{m, y_5^2+1, x_{10}+y_5\}$. It shows that the underlying structure cannot be a circle packing in the unit square, since then y_5^2+1 cannot be zero.

Let us now calculate the correct minimal polynomial for 10 circles.

Example 11.15. Find the $P_{10}(m)$ minimal polynomial.

The quadratic polynomial system of equations associated with the $n=10$ circle packing problem is (see also Figure 13.2 in the Appendix):

$$
\begin{array}{rr}
(x_1 - x_2)^2 + (y_1 - y_2)^2 = m^2 & (x_1 - x_4)^2 + (y_1 - y_4)^2 = m^2 \\
(x_2 - x_3)^2 + (y_2 - y_3)^2 = m^2 & (x_2 - x_5)^2 + (y_2 - y_5)^2 = m^2 \\
(x_5 - x_6)^2 + (y_5 - y_6)^2 = m^2 & (x_3 - x_6)^2 + (y_3 - y_6)^2 = m^2 \\
(x_4 - x_7)^2 + (y_4 - y_7)^2 = m^2 & (x_5 - x_7)^2 + (y_5 - y_7)^2 = m^2 \\
(x_7 - x_9)^2 + (y_7 - y_9)^2 = m^2 & (x_7 - x_{10})^2 + (y_7 - y_{10})^2 = m^2 \\
(x_8 - x_{10})^2 + (y_8 - y_{10})^2 = m^2 & (x_6 - x_8)^2 + (y_6 - y_8)^2 = m^2.
\end{array}
$$

Applying the simplification steps based on obvious geometrical properties of the problem, the quadratic system of equations can be reduced to:

$$
\begin{aligned}
x_2^2 + y_1^2 &= m^2, \\
x_{10}^2 + (1 - y_1 - m)^2 &= (2m)^2, \\
(1 - x_{10})^2 + (1 - y_5 - m)^2 &= m^2, \\
(1 - x_2 - m)^2 + y_5^2 &= m^2, \\
(2 - 2m - x_{10})^2 + (2y_5 - 1 - y_1 - m)^2 &= (2m)^2.
\end{aligned}
$$

Take the terms of the system of equations to the left-hand side of the equations, and set the right-hand side to zero. Then the minimal polynomial in the Gröbner basis generated by the polynomials of the left-hand side can be obtained in the following way by using the portion of Maple devoted to the generation of Gröbner bases:

```
>with(Groebner):univpoly(m, [polynomials],{x2,y1,x10,y5,m});
```

Here the previously described polynomials must be listed separated by commas in the place of the word `polynomials`. The detailed solution to this will be given in the next subsection.

11.1.3 Finding minimal polynomials using Maple

The minimal polynomial of a point arrangement problem can even be guessed with the help of a symbolic algebra system like Maple if a large enough number of exact digits are known for the m_n value. First we apply the

```
>Digits:=a;
>with(PolynomialTools):MinimalPolynomial(m,b);
```

procedure of Maple, where a is the number of exact decimal digits, and b is the degree of the minimal polynomial we can interpolate with. Table 11.2 shows the precise values which Maple requires to be able to provide the sort of $P_n(m)$ minimal polynomial we are looking for. However, in order to use this method we must have the guaranteed reliability digits of m_n, so the results of interval arithmetic based calculation schemes or the respective checking are of great importance here.

The list of minimal polynomials determined so far is

n	degree	precision	n	degree	precision
2	2	3	18	2	10
3	4	10	19	10	58
4	1	3	20	2	10
5	2	4	23	4	10
6	2	9	24	4	10
7	2	6	25	1	4
8	4	5	27	2	15
9	1	3	30	2	13
10	18	193	34	4	10
11	8	20	35	4	13
12	2	11	36	1	4
13	40	1217	39	2	13
14	2	7	42	2	13
15	4	7	52	2	14
16	1	4	56	2	14
17	8	19	99	2	17

Table 11.2. The necessary precision for determining the minimal polynomials.

$$
\begin{array}{ll}
n=2 & m^2-2 \\
n=3 & m^4-16m^2+16 \\
n=4 & m-1 \\
n=5 & 2m^2-1 \\
n=6 & 36m^2-13 \\
n=7 & m^2-8m+4 \\
n=8 & m^4-4m^2+1 \\
n=9 & 2m-1 \\
n=10 & 1180129m^{18}-11436428m^{17}+98015844m^{16}-462103584m^{15} \\
& +1145811528m^{14}-1398966480m^{13}+227573920m^{12}+1526909568m^{11} \\
& -1038261808m^{10}-2960321792m^{9}+7803109440m^{8}-9722063488m^{7} \\
& +7918461504m^{6}-4564076288m^{5}+1899131648m^{4}-563649536m^{3} \\
& +114038784m^{2}-14172160m+819200
\end{array}
$$

$$
\begin{array}{ll}
n=11 & m^8+8m^7-22m^6+20m^5+18m^4-24m^3-24m^2+32m-8 \\
n=12 & 225m^2-34 \\
n=13 & 53228084201719249374 09m^{40}+5867739593380498861732 32m^{39}
\end{array}
$$

$$
\begin{array}{l}
+13024448845332271203266928m^{38}-1298840956705690999017 0432m^{37} \\
-66972175395892949739372512m^{36}-271451157211281654252175360m^{35} \\
+143832234297958507613974 2976m^{34}-335429895467663916497996800m^{33} \\
-654369925972684882159221 6832m^{32}+944137136101134536216646 8608m^{31} \\
+101821806026335013972322549 76m^{30}-42246019864541071922661621760m^{29} \\
+37620100408876038921186476032m^{28}+28699095956807539331396009984m^{27} \\
-10258760829364534641100495257 6m^{26}+103509313296807875445571190784m^{25} \\
-23909360523055293307841740800m^{24}-62735581440162634955836358656m^{23} \\
+88454871551963142041952583680m^{22}-53012494559549527012040245248m^{21}
\end{array}
$$

	$+2135173605242212884072628224m^{20}+26378985900767549703436894208m^{19}$
	$-26497225761631816480192462848m^{18}+12731474183761933022491836416m^{17}$
	$-398432339928038268662185984m^{16}-4422001291286852186186711040m^{15}$
	$+3658751900977247115934695424m^{14}-1429726216634427968279543808m^{13}$
	$+57770773621828718826618880m^{12}+275582370688699861317976064m^{11}$
	$-171632310725283375512289280m^{10}+46974915155899860050247680m^{9}$
	$+1760067432596599241441280m^{8}-7491112055212411797372928m^{7}$
	$+3652998504696614282592256m^{6}-1072642406499215430647808m^{5}$
	$+217086289997205686190080m^{4}-30811405631471617048576m^{3}$
	$+2960075719794736758784m^{2}-174103532094609162240m$
	$+4756927106410086400$
$n=14$	$13m^2-16m+4$
$n=15$	$2m^4-4m^3-2m^2+4m-1$
$n=16$	$3m-1$
$n=17$	$m^8-4m^7+6m^6-14m^5+22m^4-20m^3+36m^2-26m+5$
$n=18$	$144m^2-13$
$n=19$	$242m^{10}-1430m^9-8109m^8+58704m^7-78452m^6$ $-2918m^5+43315m^4+39812m^3-53516m^2+20592m$ -2704
$n=20$	$128m^2-96m+17$
$n=23$	$16m^4-16m^2+1$
$n=24$	$m^4-16m^3+20m^2-8m+1$
$n=25$	$4m-1$
$n=27$	$1600m^2-89$
$n=30$	$1202m^2-252m+13$
$n=34$	$m^4+28m^3-10m^2-4m+1$
$n=35$	$46m^4-84m^3+50m^2-12m+1$
$n=36$	$5m-1$
$n=39$	$1732m^2-68m-17$
$n=42$	$864m^2-360m+37$
$n=52$	$7056m^2-193$
$n=56$	$1715m^2-588m+50$
$n=99$	$28900m^2-389$

For the $n=19$ problem instance we can find a 14 degree minimal polynomial in [130]. Although the result of the authors is correct, they may have failed to notice that the given polynomial can be divided by $m^2(2m^2+2m-1)$ without a remainder term. After doing this we get the 10 degree minimal polynomial listed above.

11.1.4 Determining exact values

Once we have the minimal polynomials, the parameters of the circle packings can be determined to arbitrary precision even if a numerical procedure is employed. With the help of minimal polynomials, the exact values of these parameters of circle packings or point arrangements can also found and expressed in terms of square root factors.

Of course this is not surprising, because if the polynomial is at most of degree 4, then there are known expressions for calculating the roots. It is interesting, however that we can determine the roots even for minimal polynomials of a degree larger than 4 by algebraic methods, and the indispensable help is provided by knowing the structure of the circle packing. Table 11.3 gives the calculated exact values for up to 100 circles.

We will discuss the determination of the exact value of $\overline{m}_{11}$, since the symbolic solution of the $P_{11}(m) = 0$ equation was only possible with a computer algebra system. It was devised in such a way that we could use the fact that the function in the given equation is the minimal polynomial of a point arrangement.

11.1.5 Solving the $P_{11}(m) = 0$ equation by radicals

The equation $m^8 + 8m^7 - 22m^6 + 20m^5 + 18m^4 - 24m^3 - 24m^2 + 32m - 8 = 0$ belongs to the class of algebraic equations with degree higher than 4 that can be solved using square root signs. However, finding the solution is difficult, even with a symbolic algebraic program. For example, when we make a simple attempt with the `roots` solver of Maple, we obtain no symbolic solution. Despite this, the equation can be solved, and we can do this with Maple. When doing the latter we have to tell the program in which quadratic field it should search for roots. In the case of the present equation, it is the quadratic field of

$$\mathbb{Q}\left[\sqrt{2}, \sqrt{3}, \sqrt{1+2\sqrt{2}}\right].$$

When we can guess that the solution must be in this field, Maple is capable of returning the exact root.

Of course it is natural to ask how we are supposed to know which quadratic field is the one where the root lies—without knowing the solution itself. In the case of the symbolic solution of the equation $P_{11}(m) = 0$, this information can be provided by the structure of the related circle packing. The idea was to study which radicals appear when we try to describe the distances between the points with the help of $\overline{m}_{11}$, and then the field of rational numbers is extended by the generated radicals.

The roots applied in the earlier field extension can be found in the following way using the notation of Figure 11.5:

a) $\overline{P_2P_3} = \sqrt{2}\overline{m}_{11}$,

b) $\overline{P_4P_6} = \sqrt{3}\overline{m}_{11}$,

c) The height of the trapezoid $P_2P_3P_4P_5$ is $\frac{1}{2}\sqrt{1+2\sqrt{2}}\overline{m}_{11}$.

11.2 Classifying circle packings based on minimal polynomials

The pattern classes discussed in Chapter 10 were defined in such a way that the packing structures in a class were similar, and in this way these circle packings could

n	r_n	m_n
2	$\frac{1}{2}(2-\sqrt{2})$	$\sqrt{2}$
3	$\frac{1}{2}(8-5\sqrt{2}+4\sqrt{3}-3\sqrt{6})$	$\sqrt{6}-\sqrt{2}$
4	$\frac{1}{4}$	1
5	$\frac{1}{2}(-1+\sqrt{2})$	$\frac{1}{2}\sqrt{2}$
6	$\frac{1}{46}(-13+6\sqrt{13})$	$\frac{1}{6}\sqrt{13}$
7	$\frac{1}{13}(4-\sqrt{3})$	$4-2\sqrt{3}$
8	$\frac{1}{4}(1+\sqrt{2}-\sqrt{3})$	$\frac{1}{2}(\sqrt{6}-\sqrt{2})$
9	$\frac{1}{6}$	$\frac{1}{2}\sqrt{2}$
11	(see below)	(see below)
12	$\frac{1}{382}(-34+15\sqrt{34})$	$\frac{1}{15}\sqrt{34}$
14	$\frac{1}{33}(6-\sqrt{3})$	$\frac{2}{13}(4-\sqrt{3})$
15	$\frac{1}{4}(8-5\sqrt{2}+4\sqrt{3}-3\sqrt{6})$	$\frac{1}{2}(1+\sqrt{2}-\sqrt{3})$
16	$\frac{1}{8}$	$\frac{1}{3}$
18	$\frac{1}{262}(-13+12\sqrt{13})$	$\frac{1}{12}\sqrt{13}$
20	$\frac{1}{482}(65-8\sqrt{2})$	$\frac{1}{16}(6-\sqrt{2})$
23	$\frac{1}{2}(-7-5\sqrt{2}+4\sqrt{3}+3\sqrt{6})$	$\frac{1}{4}(\sqrt{6}-\sqrt{2})$
24	$\frac{1}{92}(21-5\sqrt{2}+3\sqrt{3}-4\sqrt{6})$	$\frac{1}{2}(8-5\sqrt{2}+4\sqrt{3}-3\sqrt{6})$
25	$\frac{1}{10}$	$\frac{1}{4}$
27	$\frac{1}{3022}(-89+40\sqrt{89})$	$\frac{1}{40}\sqrt{89}$
30	$\frac{1}{1202}(126-5\sqrt{10})$	$\frac{1}{75}(20-\sqrt{10})$
34	$\frac{1}{4}(\sqrt{6}-\sqrt{2})$	$-7-5\sqrt{2}+4\sqrt{3}+3\sqrt{6}$
35	(see below)	$\frac{1}{46}(21-5\sqrt{2}+3\sqrt{3}-4\sqrt{6})$
36	$\frac{1}{12}$	$\frac{1}{5}$
39	$\frac{1}{866}(-17+15\sqrt{34})$	$\frac{1}{764}(-34+15\sqrt{34})$
42	$\frac{1}{2522}(217-12\sqrt{3})$	$\frac{1}{72}(15-\sqrt{3})$
52	$\frac{1}{13726}(-193+84\sqrt{193})$	$\frac{1}{84}\sqrt{193}$
56	$\frac{1}{4706}(344-7\sqrt{14})$	$\frac{1}{245}(42-\sqrt{14})$
99	$\frac{1}{57022}(170\sqrt{389}-389)$	$\frac{\sqrt{389}}{170}$

$$r_{11} = \frac{1}{568} \quad 176 - 9\sqrt{2} - 14\sqrt{3} - 13\sqrt{6}$$
$$-2\sqrt{-16523 + 12545\sqrt{2} - 9919\sqrt{3} + 6587\sqrt{6}}$$

$$m_{11} = \frac{1}{4} \quad -4 - 3\sqrt{2} + 2\sqrt{3} + 3\sqrt{6} + \quad 4 + \sqrt{2} - 2\sqrt{3} - \sqrt{6} \quad \sqrt{1 + 2\sqrt{2}}$$

$$r_{35} = \frac{1}{772} \quad 112 - 17\sqrt{2} + 8\sqrt{3} - 15\sqrt{6}$$

Table 11.3. The exact r_n and m_n parameter values for circle packing problems.

be described in an algorithmic way. The number of circles in a pattern class have a number theoretical relationship that could be used for identification of the class.

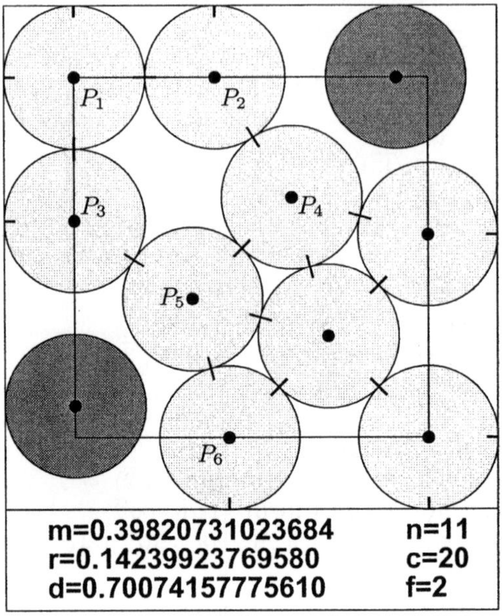

Fig. 11.5. The optimal packing for eleven circles in the square.

Of course we can ask how we are supposed to classify the remaining circle packings, the structures of which are more difficult to recognize. The minimal polynomials can help us here as well. The idea is to have those packings in the same class that have an identical type of minimal polynomials. When we do this we once again encounter pattern classes that were defined in previous chapters. The following list concerns the first 100 circle packings, and obviously only the presently known cases are discussed.

11.2.1 The linear class

Proven linear ($P_n(m) = am + b$) minimal polynomials have the $n = 4, 9, 16$, and 25 circle packings, and it is conjectured that this is also true for the $n = 36$ case. These are the precise packings of the **PAT**(k^2) pattern class. The general form for their minimal polynomials is thus

$$P_n(m) = (k-1)m - 1.$$

11.2.2 The quadratic class

The quadratic class contains those packings which have quadratic minimal polynomials. Here we distinguish two subclasses based on whether they have a linear term or not.

1. subclass

The first subclass contains the type of packings that have a quadratic minimal polynomial of the form $P_n(m) = am^2 + b$, i.e. the coefficient of the linear term is zero.

Such packings include those for $n = 5, 18$, and 27, and most probably those with 39 and 52 circles, which are precisely the packings of the $\mathbf{PAT}(k^2 + \lfloor \frac{k}{2} \rfloor)$ pattern class. The corresponding minimal polynomials have the general form

$$P_n(m) = 4k^2(k-1)m^2 - 5k^2 + 8k - 4.$$

On the other hand such packings are of course those that have $n = 2, 6$, and 12 circles, i.e. precisely those in the $\mathbf{PAT}_1(k(k+1))$ pattern class. The related minimal polynomials have the form

$$P_n(m) = k^2(2k-1)^2m^2 - 5k^2 + 4k - 1.$$

Most likely the $n = 99$ case belongs to this subclass as well.

2. subclass

These are the packings with a quadratical minimal polynomial of the form $P_n(m) = am^2 + bm + c$, where a, b, and c are all non-zero coefficients.

These kinds of packings are the optimal packings for $n = 7, 14$ circles, that is the elements of the pattern class $\mathbf{PAT}_1(k^2 - 2)$. The general form of these minimal polynomials is

$$P_n(m) = (4k^2 - 16k + 13)m^2 - (8k - 16)m + 4.$$

However, it has been proved that these kinds of packings are the cases $n = 20, 30$, and probably also those for 42 and 56, that is the packings in class $\mathbf{PAT}_2(k(k+1))$. The general form of these minimal polynomials is

$$P_n(m) = k^3(k-2)m^2 - 2k^2(k-1)m + k^2 + 1.$$

11.2.3 The quartic class

Packings in the quartic class are those for which the minimal polynomial of the packing is quartic. Similar to the quadratic case, we also describe two subclasses that depend on whether a polynomial contains cubic and linear terms.

1. subclass

Here those packings are listed for which the minimal polynomial of the packing takes the following form (the cubic and linear terms are absent):

$$am^4 + cm^2 + e.$$

This will be the case for $n = 3, 8$, and 23 circles.

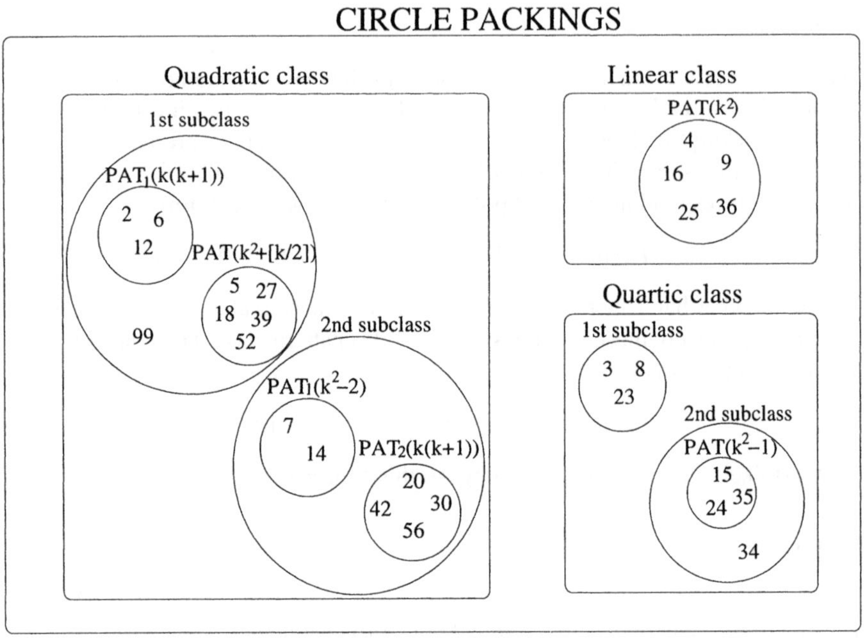

Fig. 11.6. A classification of circle packings.

2. subclass

This time the minimal polynomials take the following form:

$$am^4 + bm^3 + cm^2 + dm + e,$$

where a, b, c, d, and e are all non-zero coefficients. Proven cases with this form of minimal polynomial are the circle packings for $n = 15$ and 24, and probably also the $n = 35$ case. In other words, they are the elements of the pattern class **PAT**($k^2 - 1$) for $k = 4, 5, 6$. Then the general form for the minimal polynomial is:

$$P_n(m) = (k^4 - 12k^3 + 50k^2 - 84k + 46)m^4 + 4(-k^3 + 9k^2 - 25k + 21)m^3$$
$$+2(3k^2 - 18k + 25)m^2 - 4(k-3)m + 1.$$

Most likely this is true for the $n = 34$ case as well.

In Figure 11.6 are shown the currently known classes and their corresponding circle packings.

11.2.4 Some classes with higher degree polynomials

Among the known minimal polynomials for circle packings there are two polynomials of degree 8 (for the problem instances $n = 11$ and 17), and one polynomial of degree

10 ($n = 19$), and one of degree 18 ($n = 10$) and also one of degree 40 ($n = 13$). Other new pattern classes could of course be found by further investigations of the general problem.

12

About the Codes Used

The CD attached to the book contains the programs explained below. It is a live-CD. This means that independent of the operation system installed on the PC, you can boot your computer from this CD. The boot sequence must begin with the CD drive. Be patient, after a few minutes the Knoppix Linux distribution will appear together with a short explanatory users' guide.

12.1 The threshold accepting approach

Here the graphical user interface (GUI) and the parameters of the TAMSASS-PECS algorithm are described. This software package was developed under Linux using a GNU C compiler. This GUI needs the packages Tcl/Tk to run.

The GUI has two different windows. The first is the main window where all the parameters for the TAMSASS-PECS algorithm can be set to make the algorithm run. An optional window can also be used where the user can see graphically how the program performs a packing (see Figure 12.2). Let us call the latter window the graphical output window (GOW) and the former the main window (MW).

In the MW, the user can get some help about each option. One just has to select the option and press the F1 key. But here we will describe the possible parameters and their possible values. The mandatory parameters are the number of **Circles** (the default value is 10) and the desired accuracy, given by the variable **Epsilon** (by default it is 10^{-3}). The optional parameters are the following:

Graphical output: By default it is enabled and allows the user to see graphically how the algorithm proceeds. It does not have any associated value.

Reduce speed: It only works when the **Graphical Output** is active and the GOW appears. By default, there is no speed reduction. If the user wants to see the selected circle centre and its random movement in the GOW, the algorithm's running speed sometimes has to be reduced.

A value of 1000 means here that the delay is one second. In Figure 12.2 a snapshot of a run can be seen where the selected centre is large. In the actual algorithm run

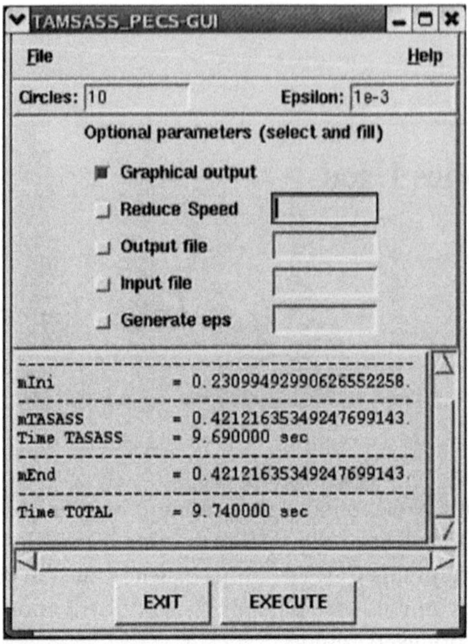

Fig. 12.1. The TAMSASS-PECS main window (MW) where the algorithm parameter can be set.

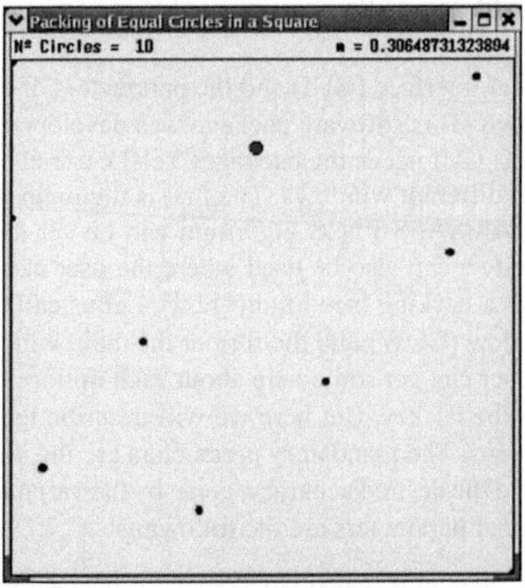

Fig. 12.2. TAMSASS-PECS running in a graphical output Window (GOW).

the selected centre is red and the other centres are blue. The centres of the circles have a width proportional to their current circle diameter.
The other information in the window contains the number of circles and the current m value of the solution. Once the algorithm has completed its task, the

user will obtain a graph similar to those presented in Figures 13.1 to 13.34 in the Appendix.

Output file: When selected, the user has to assign a name to the output file where the positions of the circle centres for the particular packing are saved. An example of an output file is the following:

```
**************************************************
                  Packing 10 circles
**************************************************
MinDistance 0.421279509472756608 09
             X                         Y
0 0.62300973099106737862 0.00000000000000000000
1 1.00000000000000000000 0.18802899175127260611
2 0.42112321130397328828 1.00000000000000000000
3 0.31150485141047462578 0.28362150588527257344
4 0.00000038647247816736 0.56724507653438316357
5 1.00000000000000000000 0.60930868988927966434
6 0.00000000000000000000 0.98852521308583807258
7 0.84240303032288488261 1.00000000000000000000
8 0.57872044975106862186 0.60930885458811179234
9 0.00000000000000000000 0.00000000000000000000
Time = 23.190000 sec.
```

Additional runs with the same output file will concatenate the outputs in the same file.

Input file: The user must provide an input file in the same format as that generated in the output file. Only the first packing will be considered in the file. We can use it for one of three purposes.

- To provide an initial configuration for the algorithm.
- To visualize a packing. For this purpose a large ϵ value can be used, e.g. 1. One can obtain an encapsulated postscript file containing the packing details if the **Generate eps** option is selected.
- To have a more accurate result, in which case one should reduce the previous **Epsilon** value.

Generate eps: It works only when the **Graphical output** is active. Then it generates an encapsulated postscript version of the GOW in the file specified by the user.

The MW also shows the command line to call the program and the returned information. This contains the solution for the initial random configuration, the final solution, and the CPU time used.

For those users who want to use the command line option, the program help option can be accessed by typing:

> *./packcircles* (without the obligatory parameters), or
> *./packcircles -h*

The help you obtain is the following:

```
Parameter -n is necessary. Options are:
 [ -h To show this help ],
   -n <Number of Circles>,
   -tasass  EpsilonTASass,
 [ -ds <Draw speed> (1000 = 1sc.) ],
 [ -in <Output file> ],
 [ -out <Output file> ],
 [ -eps <Output file in eps format> ],
 [ -tcl To get graphical output with "| escala.tcl" ].
```

For instance a command that runs without an input file is:

> *./packcircles –n 10 –tasass 1e-3 –eps Circles10.eps –tcl | escala.tcl*

A command line with all the parameters might be:

> *./packcircles -n 10 -tasass 1e-3 -ds 100 -in Circles10-1e-3.txt -out Circles10-1e-6 -eps Circles10-1e-6.eps -tcl | escala.tcl*

The standard output of the *packingcircles* program will be the standard input of *escala.tcl* program that produced the graphical output. Basically *packingcircles* ask *escala.tcl* to draw and delete the various elements that appear in the GOW.

12.2 Pulsating Disk Shaking

The program `csq.c` is an experimental code which was written for Linux's `g++`-compiler. Whenever possible use the GNU compiler to avoid trouble under other operating systems. The program was widely used to find many of the best-known packings in [110]. It is commented within the source file at great length to help other programmers to understand the underlying ideas for further developments. For later versions of the program refer to the website `www.packomania.com`.

12.2.1 Installation

At first the steps to build an executable file will be shortly explained. The program depends on a library of functions called `libnice.a` that allows the user to draw nice geometrical figures with PostScript output. Furthermore, it displays the development of the packing under consideration in an extra graphical X Window to watch the progress. Therefore you have to build `libnice.a` first. To do so, copy the file `nice-2.0.tgz` from the CD into your home directory, unpack and compile it with the following commands

```
cp [source] ~/nice-2.0.tgz
tar xzf nice-2.0.tgz
cd nice-2.0
make
```

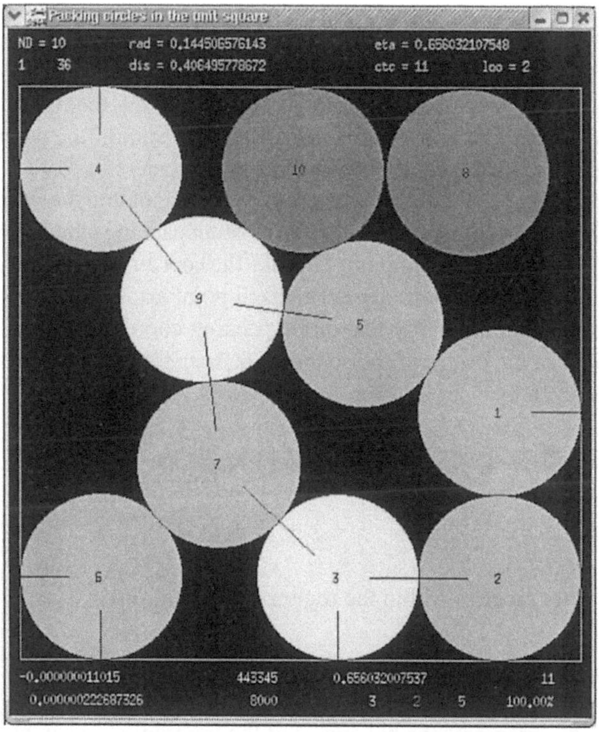

Fig. 12.3. An illustration of the Pulsating Disk Shaking algorithm.

If all things have gone right, you should end up with a file `libnice.a` in that directory. If this is not the case then some modifications in the `Makefile` are likely necessary, perhaps it would be a good idea to contact your local Unix guru. The `Makefile` can be downloaded from

`ftp://hydra.nat.uni-magdeburg.de/pub/packing/csq/Makefile`

There is a plain ASCII version, too, if the X Window System is not available. In order to build this version add the preprocessor option `-DWITHOUT_X11` as a compiler flag. Besides set the variable `DRAW` in `csq.c` to 0.

Now create a new directory `csq` below your home directory and copy the source file `csq.c` into this directory. To make an executable file, try

```
g++ -O2 -I. -I/usr/X11R6/include -I../nice-2.0 -c csq.c
g++ -o csq csq.o -L/usr/X11R6/lib -L../nice-2.0 -lnice -lX11 -lm
```

or type simply `make`.

12.2.2 Why does the program use PostScript files?

Since the result of our efforts is a geometric picture, PostScript files, being free scalable as vector graphics, are the optimal choice to draw them. Under LaTeX it is easy to include them into text documents. But where should the parameters of the packing and corresponding coordinates of the circle centers be stored? To supply an extra file is unnecessary, so they are written by the program into the PostScript files as comments. This allows the user to modify an existing packing simply by editing plain ASCII text and rerunning the program. To make the coordinates independent from the number of circles, it uses externally normalized point arrangement coordinates (i.e. $(X_i, Y_i) \in [0,1]^2$) but internally real circle packing coordinates (x_i, y_i). The latter have their origin in the centre of the square. The transformation between the sets of coordinates is given by

$$x_i = -\frac{1}{2} + r + (1-2r)X_i,$$
$$y_i = -\frac{1}{2} + r + (1-2r)Y_i$$

such that the latter range is within the region $[-\frac{1}{2}+r, \frac{1}{2}-r]^2$.

12.2.3 Usage and command-line options

As usual for Unix programs the behaviour of the program `csq` can be controlled by a couple of command-line options. To get a first overview type

```
csq -h
```

which displays the following help text:

```
Usage:  csq  -i|-r [options]  from_ND  to_ND  [SRAND]

Mandatory options: -i|-r
           -i        try to read initial configuration from an
                     existing PostScript file
           -r        set initial configuration randomly

Options:   -s   n    initial fluctuation
           -t   n    specify transformation
           -f   n    force initial improvement
           -d        derive from higher packing
           -p        read input from a pipe
           -o        only one round
           -h        display help

from_ND              number of disks to start with
to_N                 number of disks to end with
SRAND                positive integer to initialize the random
```

```
                    number generator

Please hit key 'q' in the X Window to quit.
```

Suppose you start from scratch without any PostScript file in your working directory. Then the only way to get the first results is to specify the `-r` option, i.e. initialize the circles at random:

```
csq -r 10 15 160558
```

This call runs the program to find packings for $n = 10 \ldots 15$. Note that the program normally never stops, you must hit `CTRL-C` to abort it. If the parameter `SRAND` was supplied in the command-line then the program searches for better packings. If `SRAND` is omitted the program displays only the parameters of packings calculated in previous runs.

The alternative method of the initialization is to specify the `-i` option. Then the coordinates of the circles are read in from existing PostScript files in the working directory. This method is useful if you would like to improve an existing packing. The `-s n` option tells the program the value of the fluctuation to start with. `n` should be not greater than $2 \cdot 10^{-3}$. The `-t n` option applies some initial transformations to the packing like rotations or reflections. To force the program to write a new PostScript file supply the `-f n` option. Then the value of `n` will be subtracted from the initial radius such that an "improvement" not less than `n` will be achieved when the packing is calculated. For the other options please refer to the comments within the source file. A screenshot of the program can be seen on Figure 12.3.

12.3 The interval branch-and-bound algorithm

This chapter gives a short introduction to the implementation of the interval-based optimization methods of Chapters 8 and 9. Note that earlier techniques of the present system were used to provide the numerical results of Chapter 6 as well. We introduce the version 'circpack_1.3' below. With circpack_1.3 the reader will be able to reproduce the presented results (with the exception of a few created by other tools, e.g. by Maple). For forthcoming releases of the code please refer to the

`www.sztaki.hu/~markot/packcirc.htm`

webpage.

12.3.1 Package components

As it was mentioned in Chapter 8, the optimization procedure basically consists of several optimization phases. Each phase of course depends on the results of the preceding phases. Phases 1–3 are controlled by short shell scripts. The main components of the package are the following: the branch-and-bound procedure, the polygon implementation, the tile manipulation procedures, a program for finding candidates of free circle centres, a program for verifying structural optimality and uniqueness, several utility programs, and the control scripts.

Below we briefly describe the elements of each component and summarize their functions. For further details please refer to the software documentation and to the exhaustive comments placed in the source files.

- The branch-and-bound frame

  These elements come from the C–XSC Numerical Toolbox [43] (except for the AVL tree implementation) using the interval arithmetic library PROFIL/BIAS [56], and were fine tuned to meet the demands of the circle packing problem class.

  - `AVLtree.h` contains the declaration of an AVL tree structure. We used this structure to represent `WorkList` in Algorithm 5.
  - `LIST_ari.h` offers a single-linked list structure for implementing the list `ResultList` in Algorithm 5.
  - `I_util`: in `I_util.h` has several additional utility functions which are declared for intervals.
  - `GOP.h` contains the function prototype of `AllGOP()` which actually does the optimization. Most of the work in `AllGOP()` is done by the function `GlobalOptimize()`, which is essentially the implementation of Algorithm 5. In addition, `GOP.C` contains several static functions as the main building blocks of the branch-and-bound routine (such as function evaluation and accelerating tests).
  - `GOP_ex.C` is the top-level source file of the branch-and-bound frame. Important constants like the box selection rule, the stopping tolerance, the maximal number of allowed iteration loops, and the initial best known objective function value can be set here in `main()`. Routines reading tile combinations (with their current remaining regions) and writing the output of the processed tile combinations (hopefully with smaller remaining regions) are also called here.

- The Polygon data structure

  - `Polygon.h` furnishes the polygon data structure together with those reliable polygon manipulating methods that are essential to ensure mathematical rigor for the method of active areas.

- Tile manipulation procedures

  - `Tiles.h` contains constants that determine the current tile size settings. When moving from one phase to the next one, the number of valid columns should be set here. In `Tiles.h` various tile manipulation methods like basic symmetry transformations and symmetry filtering are declared as well.
  - `GJ_utils.h` contains generic routines used by `Grow` and `Join` to manipulate sets of tile combinations.
  - `Grow.C` contains the implementation of Algorithm 10 for growing each element of a set of tile combinations by adding tiles from a new column.
  - `Join.C` contains an implementation of Algorithm 11 that joins the remaining regions of the elements of two sets of tile combinations in a pairwise fashion.

(Note that the present version of both growing and joining tile combinations are limited to our study discussed in Chapter 8. The development of a general and more flexible `Grow` and `Join` could be a subject of further studies.)

- Finding candidates of free circle centres
  - the `FreeCirc/` directory implements the algorithm proposed in Section 8.3. In particular, the source file `opt.C` provides an interface for calling the C++ version of the stochastic optimizer GLOBAL [16].
- Verifying structural optimality and uniqueness
  - Structopt.C implements the method of Chapter 9, verifying the optimality and uniqueness of the earlier conjectured packing *structures* based on our numerical results.
- Utility tools/programs
  - Binary input/output: In order to save disk space and to guarantee error-free I/O between the successive phases of the method, the tile combinations and their remaining regions are stored in binary format as blocks of storage units. `BinIO.h` offers functions for writing/reading one block to/from a file.
  - `Bin2Text`: the program compiled from `Bin2Text.C` reads a file containing the binary data of tile combinations and remaining regions and prints them in text format *applying correct outward roundings*. This promotes the following tile combinations being studied and generates the final results in a readable format.
  - `Ftildecomp`: the program compiled from `Ftildecomp.C` computes a guaranteed enclosure of the objective function over the input solution given in floating-point format. This computation is done by using long interval arithmetic. `Ftildecomp` is useful for computing the initial bounds for the optimum value and determining the function bounds over a resulting box.
  - `Startgen`: the program compiled from `Startgen.C` generates initial (full) tile combinations processed in the first phase of the method and filters out redundant (symmetric) cases.
  - `Widtheval`: the program compiled from `Widtheval.C` evaluates the approximate volume of the input box. This is useful for measuring the reduction made by the optimization method.
- Control scripts
  - `phasex.script`: the shell scripts `phase1.script` to `phase3.script` perform the operations in the corresponding phase of the method. (Note that Phase 5 consists of only one function call to `GOP_ex`, hence no control script is required for this phase.)

12.3.2 Requirements

circpack_1.3 was developed under the Linux operating system on PCs with Intel x86 architectures and compiled with the GNU C/C++ compiler gcc (version 3.2.2). circpack_1.3 was successfully built with both gcc 3.3.3 and gcc 4.0.

12.3.3 Installing and using circpack_1.3

- *Installing PROFIL/BIAS.* You have to install the interval arithmetic package PROFIL/BIAS (version 2.0) before circpack_1.3 is installed. The source archive (`Profil2.tgz`) and the documentation (`Profil2.ps.gz`, matching [56]) can be found both at their original location at

  `www.ti3.tu-harburg.de/knueppel/profil,`

  and alternatively, in the directory containing the archive of the circpack_1.3 distribution.
  - Extract `Profil2.tgz` by typing
    `tar -xvzf Profil2.tgz`
  - Add the header file `Definitions.h` to the list of "INCLUDES" in the `Profil-2.0/src/Base/Makefile` file.
  - Follow the installation instructions in `Profil.ps` to build Profil 2.0. In this particular case you have to assign the following commands:
    `cd Profil-2.0/`
    `./Configure` (enter configuration number 10 when asked)
    `gmake install`
  - (Optionally, you may assign '`gmake examples`' in order to create test examples and verify the installation.)
- *Installing circpack_1.3*
  - Extract circpack_1.3 from its archive by typing
    `tar -xvzf circpack_1.3.tgz`
  - Check the file `circpack_1.3/CONFIG` for the correct setting of the `Profil-2.0/` directory path and your compiler name.
  - Compile all programs of circpack_1.3 by simply typing
    `make all`

    Now circpack_1.3 is ready to use.

    When you modify the sources, assign '`make clear`' to delete the old intermediate and executable files and either rebuild the whole package by '`make all`', or type '`make prog`' if you want to build an individual executable called '`prog`'.
  - To test the installation, change to the `circpack_1.3/Test/` subdirectory, and type
    `./test_cp`

    This script runs some of the executables on a lots of small examples and checks the correctness of the installation.
- *Using circpack_1.3*

  The `README.ps` file contains a *data flow table* which highlights the steps of the interval methods used to prove optimality for $n = 28, 29$, and 30. This table contains the order of the individual program calls, the respective input and output files, and lists the constants and their required values. With the aid of the original study input/output files located in the `28_files/`, `29_files/`, `30_files/`, and common_files/ directories, you will be able to reproduce and study each individual step and phase of our computations.

- *Documentation*
  There is a PostScript file and also a multiple page html document available at circpack_1.3/doc. The documentation files were generated by the 'Doxygen' documentation system (www.doxygen.org).

12.4 A Java demo program

Unlike the earlier codes the present one is not too professional. It is not fast, nor does it find a solution in an efficient manner. Its merits are that it allows one to understand how such programs work. Also, due to its portability it is intended for a wider audience. The program itself is based on the algorithm by J. Donovan [5], and was coded by Viktor Váradi as a part of his Masters thesis in 2002.

The program is available at the internet address of

http://www.inf.u-szeged.hu/~csendes/pack.html

To run the program the browser must support Java programs, and the settings must be given. The start page is depicted in Figure 12.4.

How this program should be used is self-evident. First set the number of circles to be packed in the lower left slot (Number of circles), supply the number of allowed independent searches (Number of Try-outs), and push the Start button. If you simply push the start button, then the packing of nine circles into the unit square problem will be solved, and the search algorithm will be restarted ten times. For a larger number of circles to be packed it is worth checking the box "Show me only the best results". Then the right-hand square will not be updated, and it may lead to a significant program speedup. If your browser is unable to run Java programs, then you will not see the squares and the buttons. If this happens you should install the Java feature on your browser.

The program first locates a starting configuration for circles (with a smaller than optimal radius), and then it executes a relatively quick version of the stochastic optimization procedure. During a run the actual approximate solution can be seen in the square to the right, and the circles typically move according to the changes induced by the algorithm. After each independent search the best packing achieved up to that point will be displayed in the left-hand square.

Below the squares you will find some statistics on the actual run: first the density of the best packing found by the program, then the best density known from the literature (in most cases the optimal value is unknown), and finally the number of searches performed.

Setting too high a number of circles to pack may cause the program to run very slowly. In such cases you can stop the program by pushing the Stop button. Then the actual solution will remain in the squares.

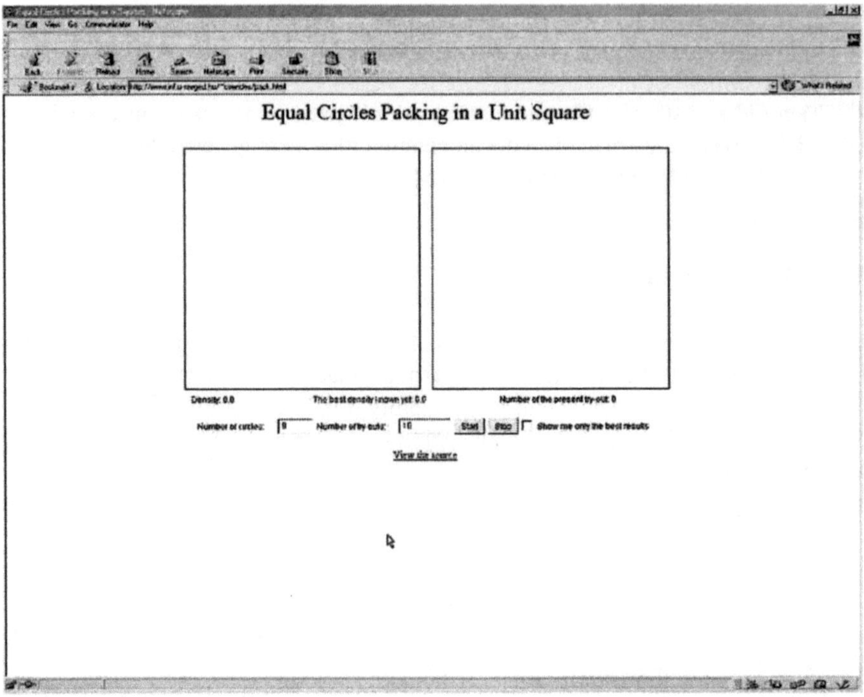

Fig. 12.4. The starting menu of the Java demonstration program for a circle packing task.

What is this program good for? Definitely not for breaking the known records, since due to the interpretive nature of the Java code it cannot compete against optimized algorithmic languages like Fortran or C. Still, it is certainly good for

- providing an impression of how such algorithms work,
- showing how easy or difficult it is to achieve a given level of suboptimality e.g. a 1% worse result than the known optimum,
- supplying good approximate solutions for users for certain problem instances,
- demonstrating what kind of computational complexity is to be expected for a similar problem,
- enticing people to this delightful field of research.

In addition to the above, a fully functional program code is now available. The interested reader should click on the "View the source" link at the bottom of the page. The full Java code is so short that it is worth copying some parts of it here. It is also easy to read even for those who are unacquainted with Java programming.

The Java method to place the starting centres of the circles randomly into the square:

```
// Places the centre points randomly into the square
```

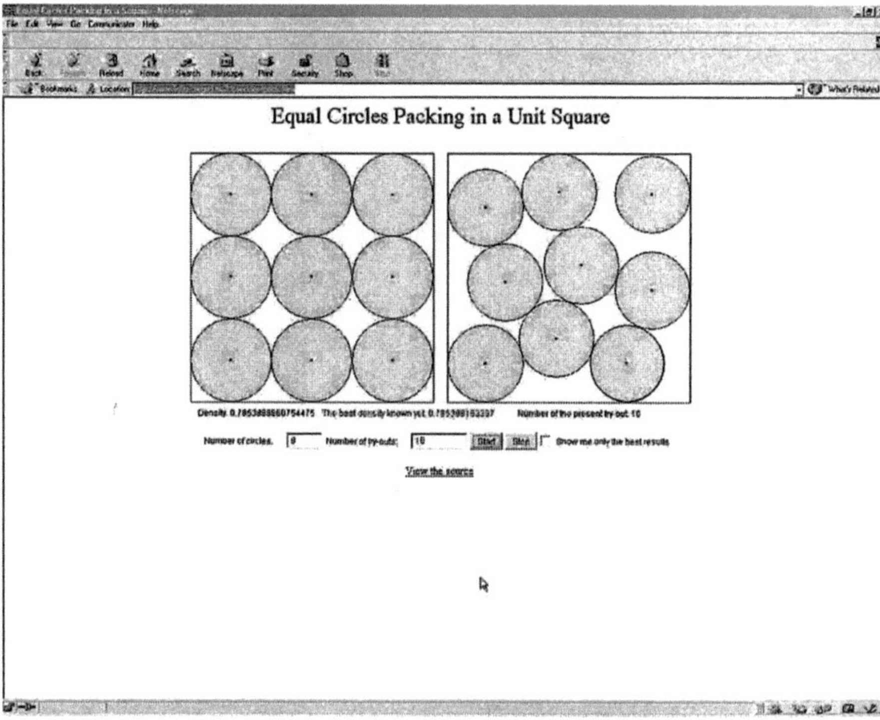

Fig. 12.5. Result for the Java demonstration program after packing 9 circles.

```
public void randomPoints() {
     double x,y;
     for(int i=0; i<=(circles.n-1); ++i) {
          x=wp*Math.random();
          y=wp*Math.random();
          origos[i] = new Coordinate(x, y);
     }
}
```

The Java method responsible for the shifting of the centre points (notice that the program code is nicely readable even without some knowledge of the programming language):

```
// Move the centres of the circles

public void movePoints() {
   boolean moved;
   double closest;
   double distance;
   double x, y;
```

```
step=200;
randomPoints();
circles.r=(minimumDistance(origos)/2.0);
if(circles.r==0.0) {
   do{
      randomPoints();
      circles.r=(minimumDistance(origos)/2.0);
   }while(circles.r==0.0);
}
do {
   if(doing==false) {
      break;
   }
   moved=false;
   for(int i=0; i<=(circles.n-1); ++i) {
      x=origos[i].x;
      y=origos[i].y;
      closest=nearestNeighbour(origos, i, x, y);
      x=origos[i].x+step;
      if(x>wp) { x=wp; }
      distance=nearestNeighbour(origos, i, x, y);
      if(distance>closest) {
        origos[i].x=x;
        moved=true;
        closest=nearestNeighbour(origos, i, x, y);
      }
      x=origos[i].x-step;
      y=origos[i].y;
      if(x<0) { x=0; }
      distance=nearestNeighbour(origos, i, x, y);
      if(distance>closest) {
        origos[i].x=x;
        moved=true;
        closest=nearestNeighbour(origos, i, x, y);
      }
      y=origos[i].y+step;
      x=origos[i].x;
      if(y>wp) { y=wp; }
      distance=nearestNeighbour(origos, i, x, y);
      if(distance>closest) {
        origos[i].y=y;
        moved=true;
        closest=nearestNeighbour(origos, i, x, y);
      }
```

```
          y=origos[i].y-step;
          x=origos[i].x;
          if(y<0) { y=0; }
          distance=nearestNeighbour(origos, i, x, y);
          if(distance>closest) {
            origos[i].y=y;
            moved=true;
          }
    }
    if(moved==false ) { step/=div; }
    circles.r=(minimumDistance(origos)/2.0);
    circlesDraw(realOrigos, 370);
    } while(step>limit);
}
```

Example 12.1. Consider the case when we want to pack just four circles. We know that the optimal packing is the regular square grid packing. Set the number of circles to be packed to 4, and let the number of tries be 10. After pushing the Start button, the result will be available within a few seconds. The density obtained is usually exactly the optimal density. In a sense this comes as no surprise, since the problem was easy, straightforward to solve and the optimal configuration has been found several times (usually in each separate search).

Example 12.2. Now set the number of circles to 10. Considering the years spent working on the solution since 1990, we might expect a less impressive result. In fact the obtained results now differ significantly in the restarted runs. The density values it finds are typically around 0.68, while the optimal density is 0.69003585264. We can also see that the unsolved problem is not just one of the fine tuning of the packing it found but even the structure of the packing can dramatically change, i.e. the problem is far from being solved. Still, it is remarkable that such a simple program with all its drawbacks of visualization and interpreting is able to supply an approximate solution within a few seconds, one that is only 1% away from the optimal one.

When increasing the number of circles to pack, the relative precision of the solution found becomes evermore worse and worse, which is due to the fact that the CPU time used for the solution is not even proportional to the number of circles, nor the exploding complexity of the problems. But it may be worthwhile experimenting with a larger number of circles to see the interesting hexagonal structures that emerge within the prospective candidate packing solutions.

13

Appendix A: Currently Best Known Results for Packing Congruent Circles in a Square

A.1. Numerical results for the packings

The following five tables summarize the numerical results for the currently best known circle packings (as of March 2006), where

r_n : is the common radius of the circles,
m_n : is the minimal distance between the points,
c_n : is the number of contacts,
f_n : is the number of free circles (loosed circles or rattlers),
d_n : is the density of the packing.

A.2. The figures of the packings

Free circles or rattlers appear in the figures with dark shading whereas circles with less than three contacts are drawn in the second darkest shading. Rigid or fixed circles have two different shadings according to their membership in 3-cliques, see Section 2 in [5]. Note that only one packing is shown for each number of circles even if more different packings exist with equal radius.

Please refer to Eckard Specht's web-site `www.packomania.com` in order to see the best packings up-to-date.

Table 13.1. Best known results for the problem of packing circles in a unit square when $2 \leq n \leq 40$.

n	r_n	m_n	c_n	f_n	d_n
2	0.292893218813	1.414213562373	5	–	0.539012084453
3	0.254333095030	1.035276180410	7	–	0.609644808741
4	0.250000000000	1.000000000000	12	–	0.785398163397
5	0.207106781187	0.707106781187	12	–	0.673765105566
6	0.187680601147	0.600925212577	13	–	0.663956909464
7	0.174457630187	0.535898384862	14	1	0.669310826841
8	0.170540688701	0.517638090205	20	–	0.730963825254
9	0.166666666667	0.500000000000	24	–	0.785398163397
10	0.148204322565	0.421279543984	21	–	0.690035785264
11	0.142399237696	0.398207310237	20	2	0.700741577756
12	0.139958844038	0.388730126323	25	–	0.738468223884
13	0.133993513499	0.366096007696	25	1	0.733264694904
14	0.129331793710	0.348915260374	32	1	0.735679255543
15	0.127166547515	0.341081377402	36	–	0.762056010927
16	0.125000000000	0.333333333333	40	–	0.785398163397
17	0.117196742783	0.306153985300	34	1	0.733550263302
18	0.115521432464	0.300462606289	38	–	0.754653357876
19	0.112265437571	0.289541991995	37	2	0.752307896742
20	0.111382347512	0.286611652352	44	–	0.779493686868
21	0.106860212352	0.271812255359	39	2	0.753357702890
22	0.105665296757	0.267958401551	43	1	0.771680112098
23	0.102802323380	0.258819045103	56	–	0.763631032126
24	0.101381800432	0.254333095030	56	–	0.774963259758
25	0.100000000000	0.250000000000	60	–	0.785398163397
26	0.096362339010	0.238734757241	56	2	0.758469090484
27	0.095420001748	0.235849528301	55	–	0.772311456467
28	0.093672833833	0.230535493643	57	1	0.771854111404
29	0.092463144040	0.226882900744	65	1	0.778906241780
30	0.091671057986	0.224502964531	65	–	0.792019026461
31	0.089338333351	0.217547291619	55	4	0.777297478729
32	0.087858157088	0.213174562590	63	3	0.776004124475
33	0.087230014136	0.211328384143	65	1	0.788852303934
34	0.085270344351	0.205604646760	80	–	0.776649064332
35	0.084290712122	0.202763600863	80	–	0.781227212999
36	0.083333333333	0.200000000000	84	–	0.785398163397
37	0.082089766429	0.196429184630	77	–	0.783302723724
38	0.081709776125	0.195342304127	77	–	0.797042556835
39	0.081367527047	0.194365063162	80	–	0.811179027292
40	0.079186752517	0.188175522018	85	2	0.787979518878

Table 13.2. Best known results for the problem of packing circles in a unit square when $41 \leq n \leq 80$.

n	r_n	m_n	c_n	f_n	d_n
41	0.078450210117	0.186099511848	100	1	0.792723899305
42	0.077801502898	0.184277072117	90	–	0.798684278653
43	0.076339810580	0.180191135457	85	1	0.787264166427
44	0.075781986018	0.178639245671	82	4	0.793842807842
45	0.074727343687	0.175716314176	94	3	0.789444268495
46	0.074272199909	0.174459360872	91	1	0.797187131992
47	0.073113151271	0.171270563823	92	3	0.789293882076
48	0.072432291280	0.169405429370	111	2	0.791144033841
49	0.071692681704	0.167386076868	120	1	0.791216989527
50	0.071377103865	0.166526577344	100	1	0.800272183998
51	0.071043138115	0.165618374312	97	3	0.808656948183
52	0.070957693073	0.165386237970	105	–	0.822530842068
53	0.069947252562	0.162648077425	108	4	0.814642500051
54	0.068645540105	0.159139516307	115	2	0.799407623084
55	0.068055360559	0.157555747530	113	2	0.800271297245
56	0.067532596322	0.156156500462	119	–	0.802351729503
57	0.067004873106	0.154747406944	113	3	0.803965673846
58	0.066232983746	0.152692531391	125	2	0.799330724318
59	0.065807496904	0.151562950614	121	4	0.802698827038
60	0.065030412648	0.149505654049	129	3	0.797139156419
61	0.064666268906	0.148544126695	121	1	0.801374124593
62	0.064252183294	0.147452679810	128	2	0.804113477817
63	0.064011528205	0.146819313688	145	1	0.810973780650
64	0.063458986813	0.145367754452	129	2	0.809685038563
65	0.063203957072	0.144699014781	137	2	0.815740018110
66	0.062862256899	0.143804230796	130	3	0.819358090712
67	0.062587429542	0.143085575883	153	–	0.824515655814
68	0.062520077998	0.142909593912	148	4	0.835021788298
69	0.061383571686	0.139948181881	137	1	0.816776573752
70	0.060596693631	0.137906776651	135	5	0.807506020669
71	0.060096531352	0.136612997248	143	3	0.805576954556
72	0.059801002127	0.135849927909	180	1	0.808908302138
73	0.059366050583	0.134728725877	151	3	0.808256206863
74	0.059082376337	0.133998672690	153	3	0.811516775123
75	0.058494535281	0.132488813742	146	2	0.806198017942
76	0.058198524936	0.131730037632	191	1	0.808699956656
77	0.057852577916	0.130844544210	174	–	0.809628951523
78	0.057702476735	0.130460772895	180	1	0.815893333309
79	0.057508497795	0.129965202741	169	3	0.820806922795
80	0.057370684147	0.129613385494	193	–	0.827217888560

Table 13.3. Best known results for the problem of packing circles in a unit square when $81 \leq n \leq 120$.

n	r_n	m_n	c_n	f_n	d_n
81	0.056869921112	0.128336855974	198	3	0.823000583646
82	0.056512271650	0.127426911812	203	4	0.822714696667
83	0.056129373022	0.126454353161	209	2	0.821501474386
84	0.055856665888	0.125762702259	183	3	0.823339926925
85	0.055680181768	0.125315548584	191	3	0.827885139567
86	0.055572999407	0.125044156483	205	2	0.834403272277
87	0.054695259721	0.122826583176	175	–	0.817651999581
88	0.054406636913	0.122099298194	187	5	0.818344762893
89	0.053947040858	0.120943129628	182	5	0.813720269423
90	0.053749948307	0.120448049480	230	1	0.816861606108
91	0.053496719884	0.119812602205	178	2	0.818173810461
92	0.053317085175	0.119362266622	156	10	0.821619044629
93	0.052926433388	0.118384170617	190	2	0.818423476521
94	0.052795362727	0.118056384766	224	2	0.823131614712
95	0.052420366496	0.117119642119	222	4	0.820112788103
96	0.052275425470	0.116757999100	224	4	0.824168967917
97	0.052114795502	0.116357483969	234	2	0.827644213165
98	0.052032987703	0.116153614607	203	2	0.833553492453
99	0.051978606450	0.116018134843	199	–	0.840299937025
100	0.051401071774	0.114581352162	216	7	0.830030826636
101	0.051078356337	0.113780115213	210	6	0.827837458205
102	0.050774421024	0.113026558149	252	4	0.826114042028
103	0.050534497608	0.112432427714	257	3	0.826348041985
104	0.050348792812	0.111972995972	199	4	0.828249793523
105	0.050242948364	0.111711307651	228	2	0.832701611959
106	0.050015200417	0.111148644272	235	3	0.833028317432
107	0.049506158485	0.109893086039	212	8	0.823857512057
108	0.049243520781	0.109246395896	199	9	0.822757444787
109	0.049060052710	0.108795091242	215	4	0.824199585136
110	0.048779469321	0.108105606913	285	1	0.822274269044
111	0.048643936322	0.107772865452	221	4	0.825145003148
112	0.048445405903	0.107285822215	261	1	0.825796605619
113	0.048257588494	0.106825454649	287	1	0.826722100560
114	0.048167755522	0.106605396386	280	1	0.830935940404
115	0.047914247537	0.105984865206	282	5	0.829424877813
116	0.047775960994	0.105646663762	249	4	0.831814956319
117	0.047643526600	0.105322968504	262	2	0.834340904499
118	0.047572362680	0.105149108402	245	2	0.838960130139
119	0.047544646218	0.105081409294	267	3	0.845084379623
120	0.047529921591	0.105045446890	241	–	0.851658165016

Table 13.4. Best known results for the problem of packing circles in a unit square when $121 \leq n \leq 160$.

n	r_n	m_n	c_n	f_n	d_n
121	0.046891999625	0.103489674838	264	5	0.835858472004
122	0.046626691121	0.102843926204	275	5	0.833256858017
123	0.046348792023	0.102168342569	263	6	0.830102694507
124	0.046099752156	0.101563619705	265	6	0.827882569457
125	0.045978336542	0.101269036794	201	18	0.830168776931
126	0.045828714873	0.100906235980	240	4	0.831372731264
127	0.045663869682	0.100506798017	265	4	0.831953432089
128	0.045473271987	0.100045320075	328	–	0.831519142297
129	0.045178743652	0.099332964371	269	6	0.827194965307
130	0.045043161976	0.099005352181	267	7	0.828611515844
131	0.044887232368	0.098628813692	338	2	0.829214381767
132	0.044746370935	0.098288883554	347	1	0.830308417788
133	0.044600223986	0.097936420557	278	4	0.831142696990
134	0.044548122181	0.097810821188	259	6	0.835436557221
135	0.044428206477	0.097521855191	266	3	0.837145995837
136	0.044353299101	0.097341424866	270	3	0.840505654420
137	0.044293023546	0.097196281458	308	–	0.844386136710
138	0.044032132019	0.096568497720	326	6	0.840559358468
139	0.043945742388	0.096360776496	323	5	0.843331425575
140	0.043804395407	0.096021081672	351	5	0.843943338354
141	0.043658469169	0.095670602431	313	6	0.844317892156
142	0.043576632733	0.095474149348	341	2	0.847121206057
143	0.043530511915	0.095363464703	297	–	0.851282016761
144	0.043002745898	0.094098477643	295	12	0.836574727482
145	0.042871343605	0.093783977453	257	7	0.837244039299
146	0.042646910194	0.093247233143	296	6	0.834214763957
147	0.042439897295	0.092752617732	375	3	0.831794139020
148	0.042329005699	0.092487848751	382	2	0.833081944918
149	0.042251859906	0.092303728197	284	4	0.835656514947
150	0.042145457730	0.092049884492	327	–	0.837033195178
151	0.041976048815	0.091645968965	360	7	0.835853051836
152	0.041777096335	0.091171995116	392	3	0.833431601713
153	0.041631845466	0.090826216993	391	3	0.833091357580
154	0.041551072470	0.090634026986	375	2	0.835285743673
155	0.041438011495	0.090365125182	335	7	0.836140739056
156	0.041298479102	0.090033447069	352	9	0.835877408234
157	0.041202212729	0.089804732875	349	3	0.837318342833
158	0.041158525256	0.089700969773	358	12	0.840865569748
159	0.041114183404	0.089595672643	402	5	0.844365217382
160	0.041044595831	0.089430466355	367	3	0.846801897163

Table 13.5. Best known results for the problem of packing circles in a unit square when $161 \leq n \leq 200$.

n	r_n	m_n	c_n	f_n	d_n
161	0.041010164330	0.089348741831	328	1	0.850665398328
162	0.040797848240	0.088845072009	339	6	0.847109209950
163	0.040681201115	0.088568552416	354	5	0.847471322588
164	0.040525041364	0.088198585368	394	4	0.846136937645
165	0.040377544085	0.087849371948	418	4	0.845110732099
166	0.040237519991	0.087518059304	348	3	0.844345847114
167	0.040148423041	0.087307351007	381	1	0.845674678732
168	0.040095593140	0.087182450400	402	2	0.848501157873
169	0.039759378928	0.086388243687	409	4	0.839297174061
170	0.039585933812	0.085978984395	364	8	0.836913517436
171	0.039434198868	0.085621204987	416	10	0.835395300146
172	0.039318677249	0.085348971855	418	8	0.835364696522
173	0.039236445096	0.085155270372	429	4	0.836710619012
174	0.039179543023	0.085021275487	441	6	0.839107984106
175	0.039046149820	0.084707286435	399	6	0.838193626231
176	0.038988813665	0.084572380933	422	7	0.840509414341
177	0.038841476832	0.084225867853	430	6	0.838908524884
178	0.038727777584	0.083958616412	455	5	0.838716189022
179	0.038682353673	0.083851883797	355	4	0.841450720093
180	0.038620111853	0.083705668246	447	–	0.843430753378
181	0.038475500327	0.083366105927	448	7	0.841776892065
182	0.038416988626	0.083228775062	444	6	0.843855137412
183	0.038324943134	0.083012808606	424	4	0.844430673944
184	0.038276473990	0.082899120002	435	6	0.846898850932
185	0.038244982284	0.082825266248	402	4	0.850101004314
186	0.038227175134	0.082783509717	396	2	0.853900424343
187	0.038200922052	0.082721953932	387	2	0.857312528370
188	0.038188623136	0.082693118985	378	–	0.861342195310
189	0.037866552704	0.081938567583	448	6	0.851379553528
190	0.037695999979	0.081539419899	421	6	0.848191685066
191	0.037544127565	0.081184237898	417	6	0.845799206937
192	0.037431098348	0.080920049346	447	3	0.845115842743
193	0.037310270862	0.080637776273	484	4	0.844041859326
194	0.037253974913	0.080506309926	484	2	0.845856786624
195	0.037213368627	0.080411503065	514	1	0.848364435455
196	0.037045076544	0.080018754887	382	5	0.845019900786
197	0.036871532689	0.079614049431	464	6	0.841392198062
198	0.036727443767	0.079278263460	502	7	0.839066665758
199	0.036660444456	0.079122198864	497	3	0.840230419966
200	0.036583075322	0.078942035506	519	2	0.840892135688

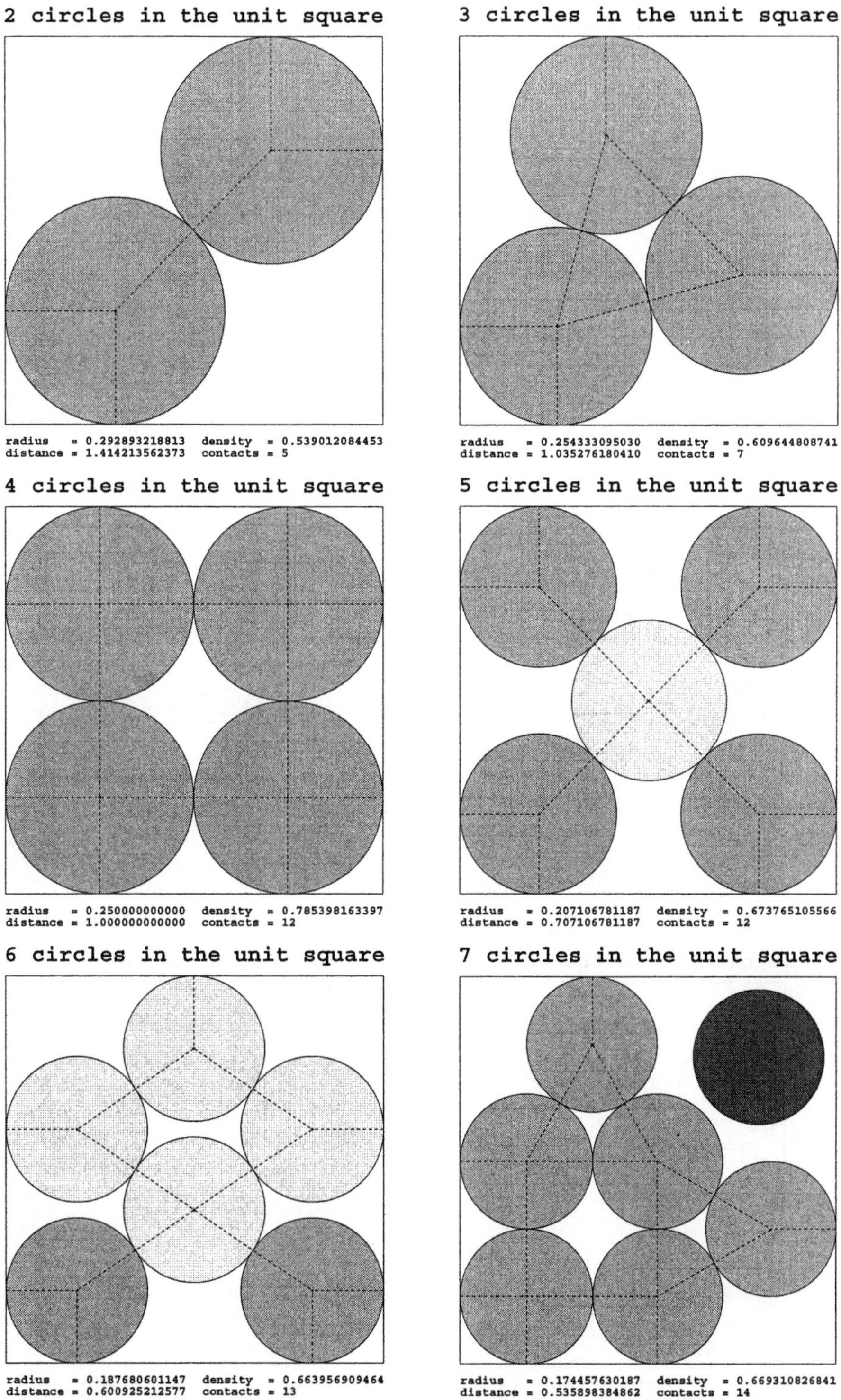

Fig. 13.1. Best known packings for 2–7 circles.

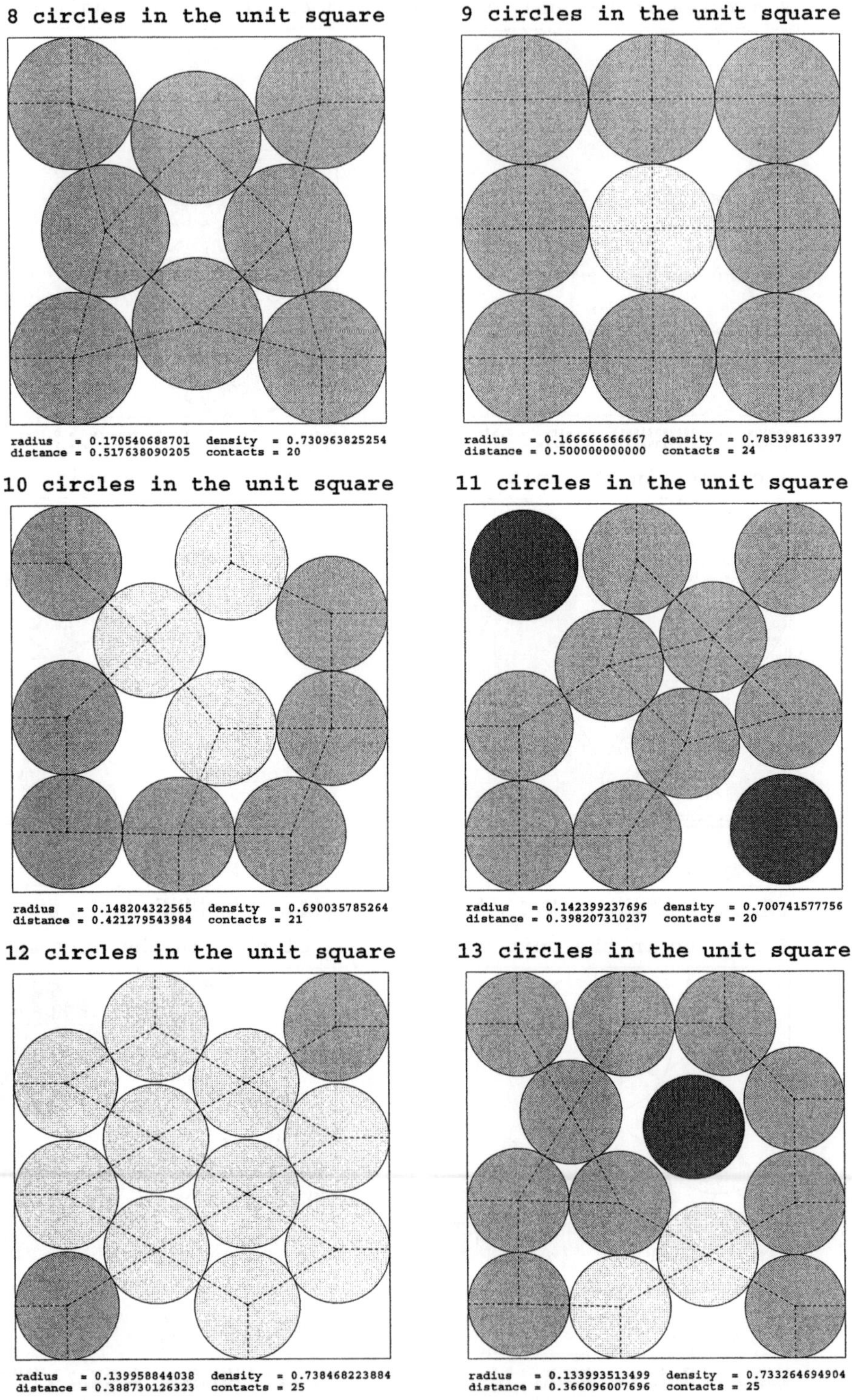

Fig. 13.2. Best known packings for 8–13 circles.

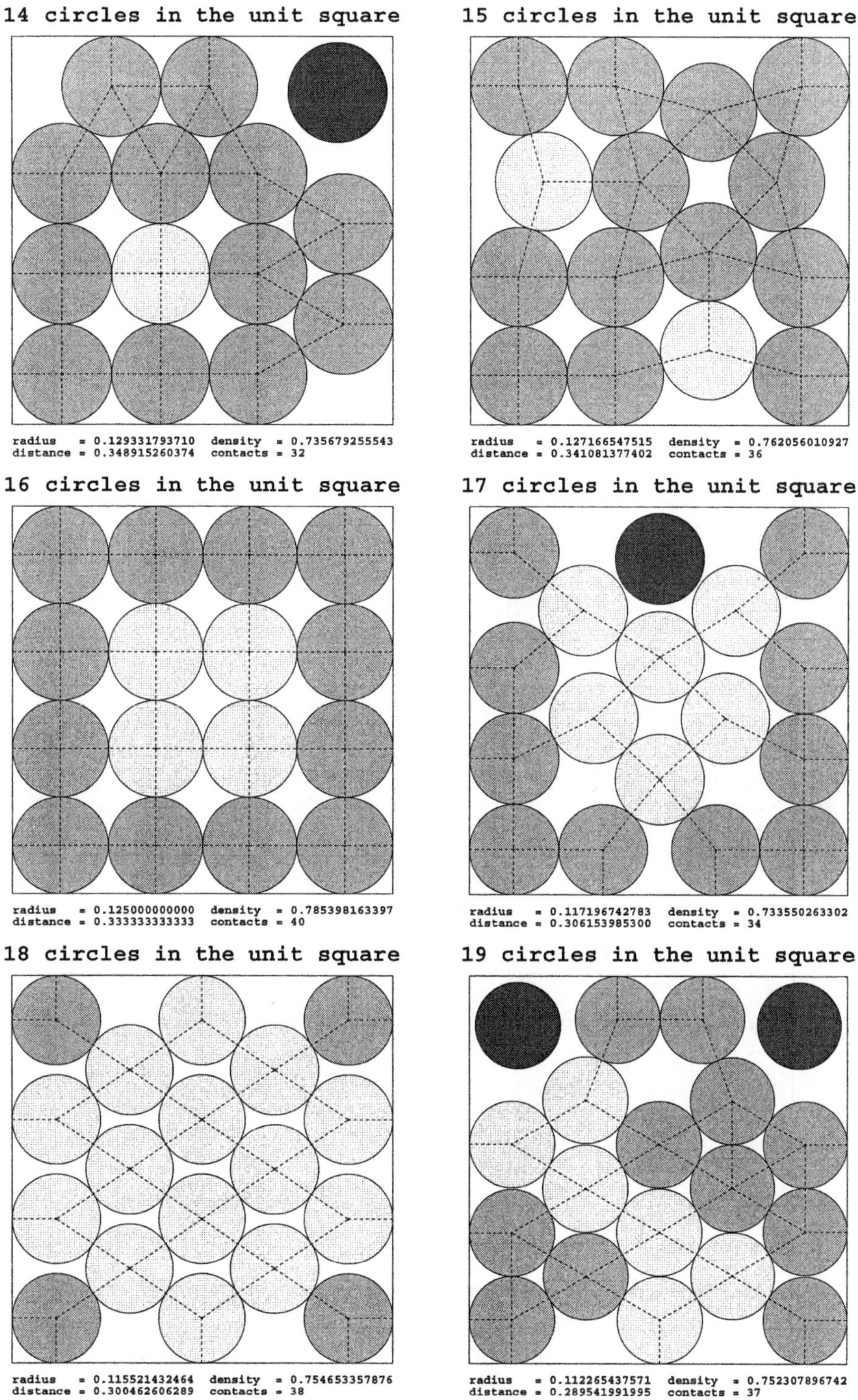

Fig. 13.3. Best known packings for 14–19 circles.

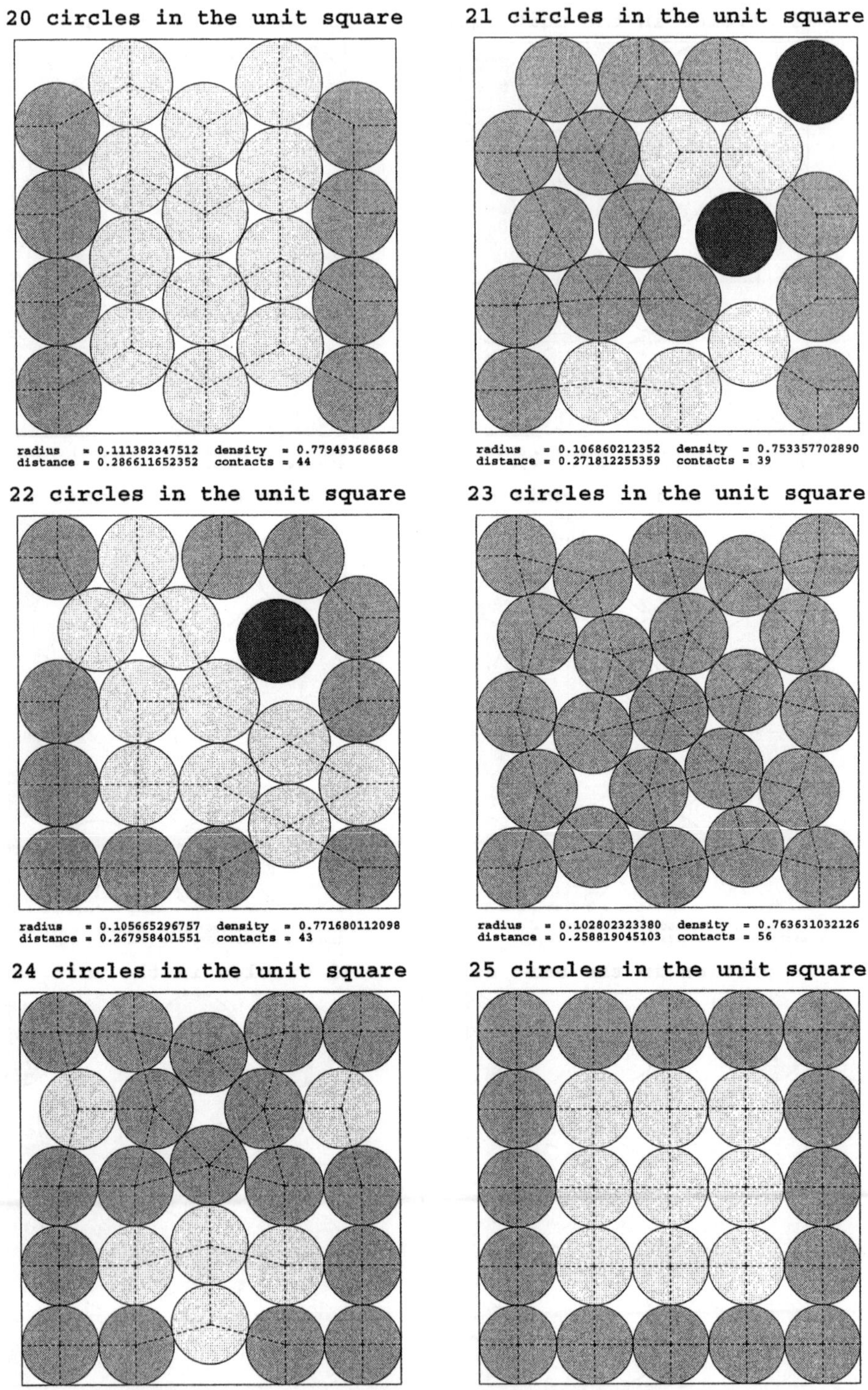

Fig. 13.4. Best known packings for 20–25 circles.

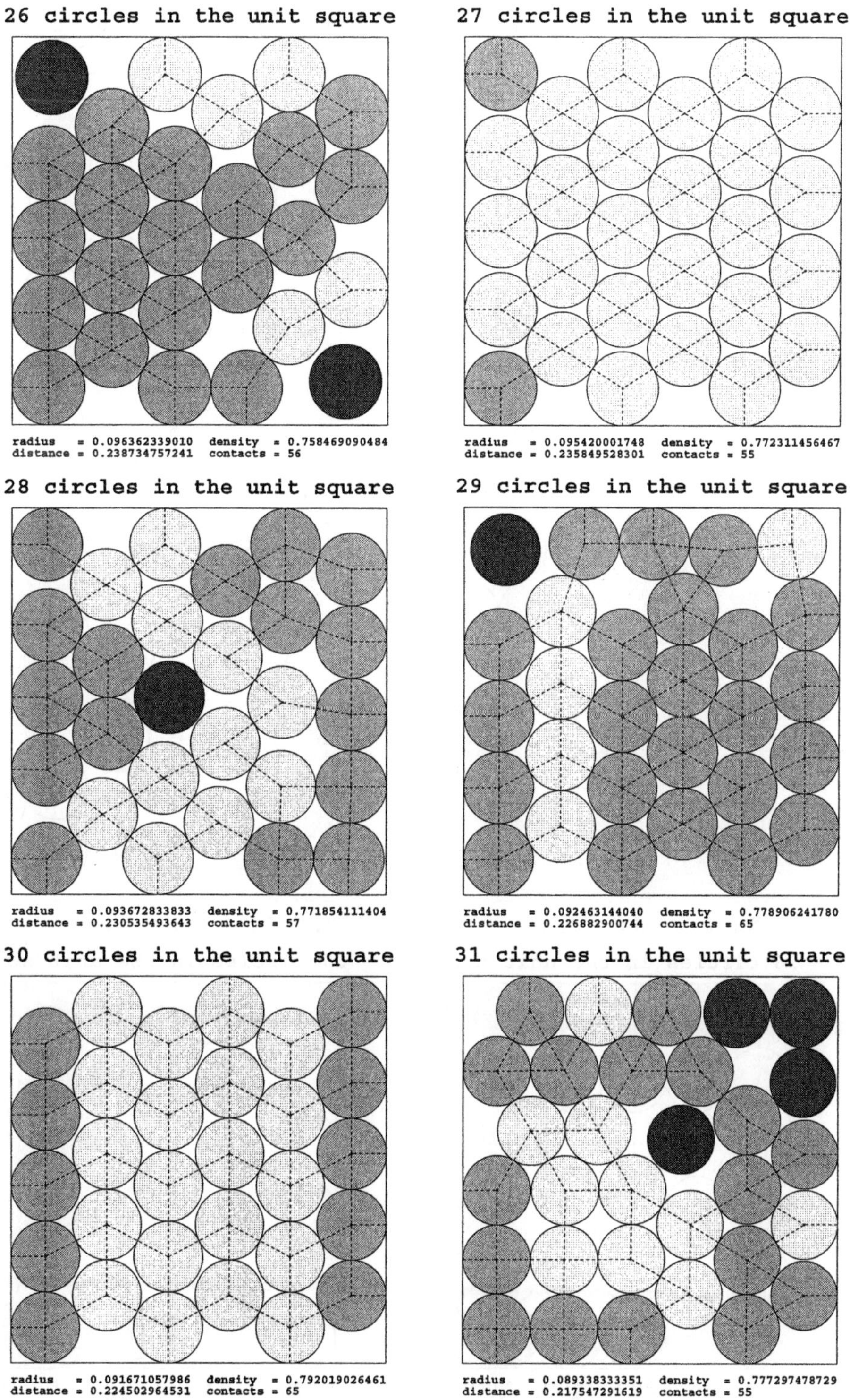

Fig. 13.5. Best known packings for 26–31 circles.

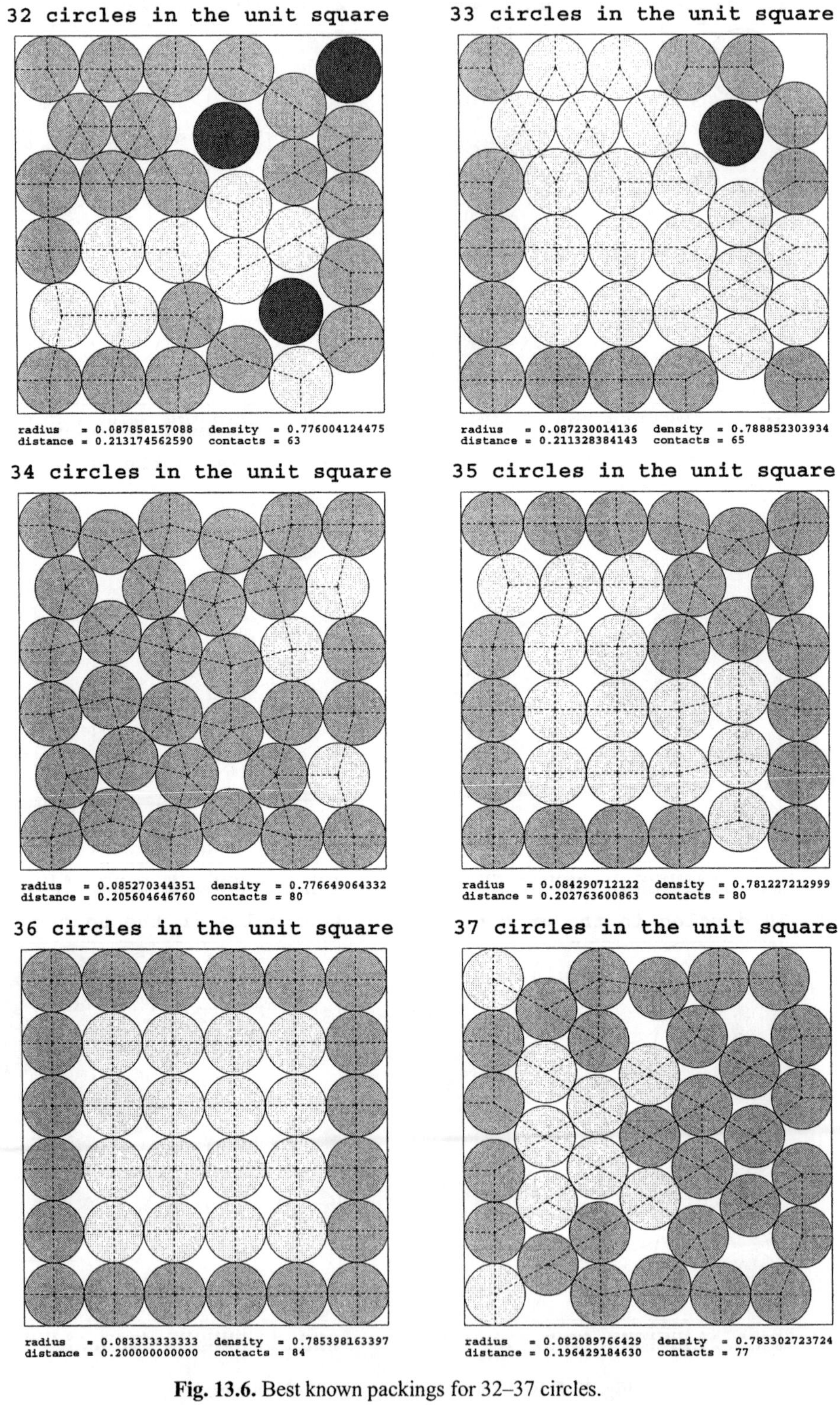

Fig. 13.6. Best known packings for 32–37 circles.

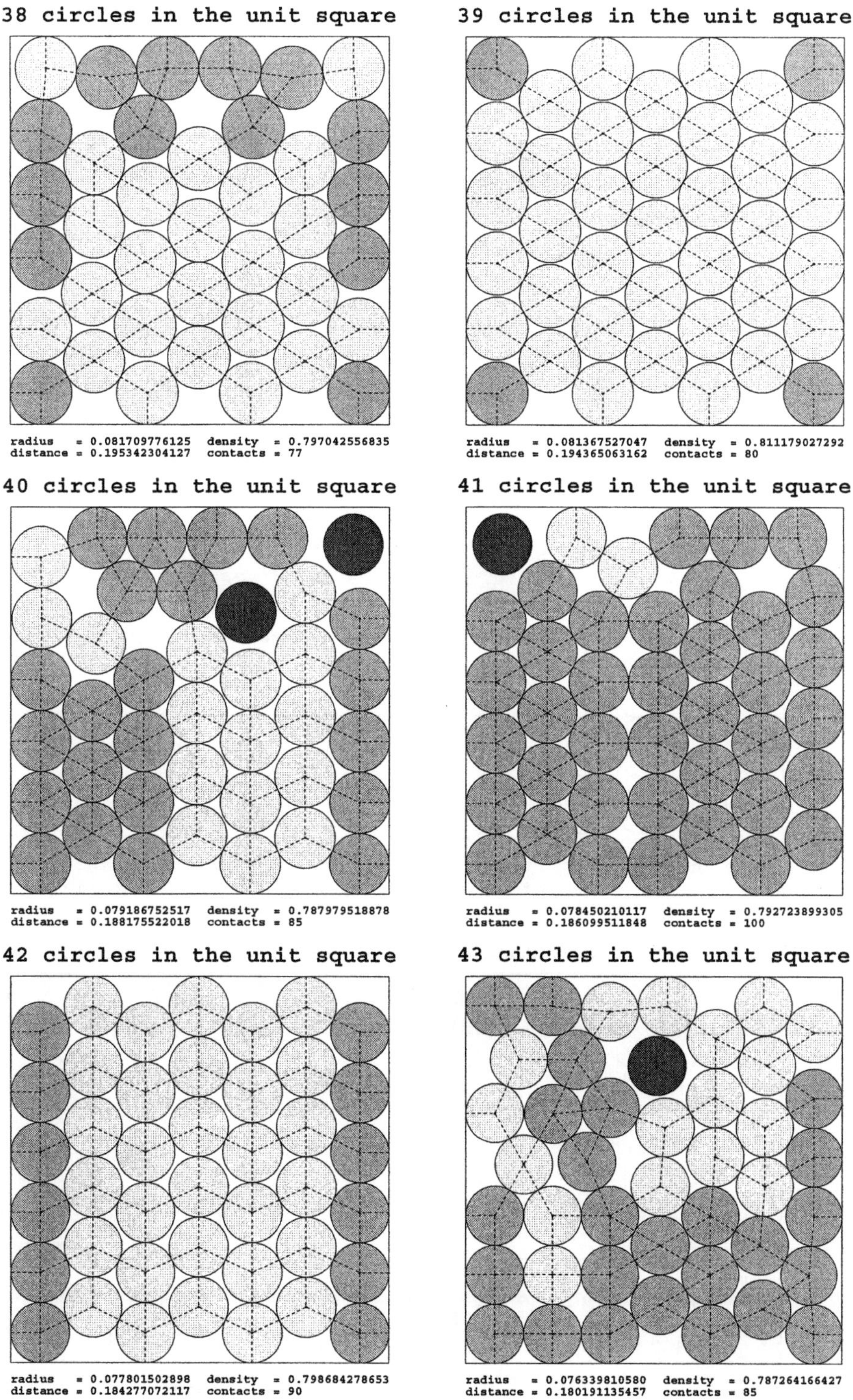

Fig. 13.7. Best known packings for 38–43 circles.

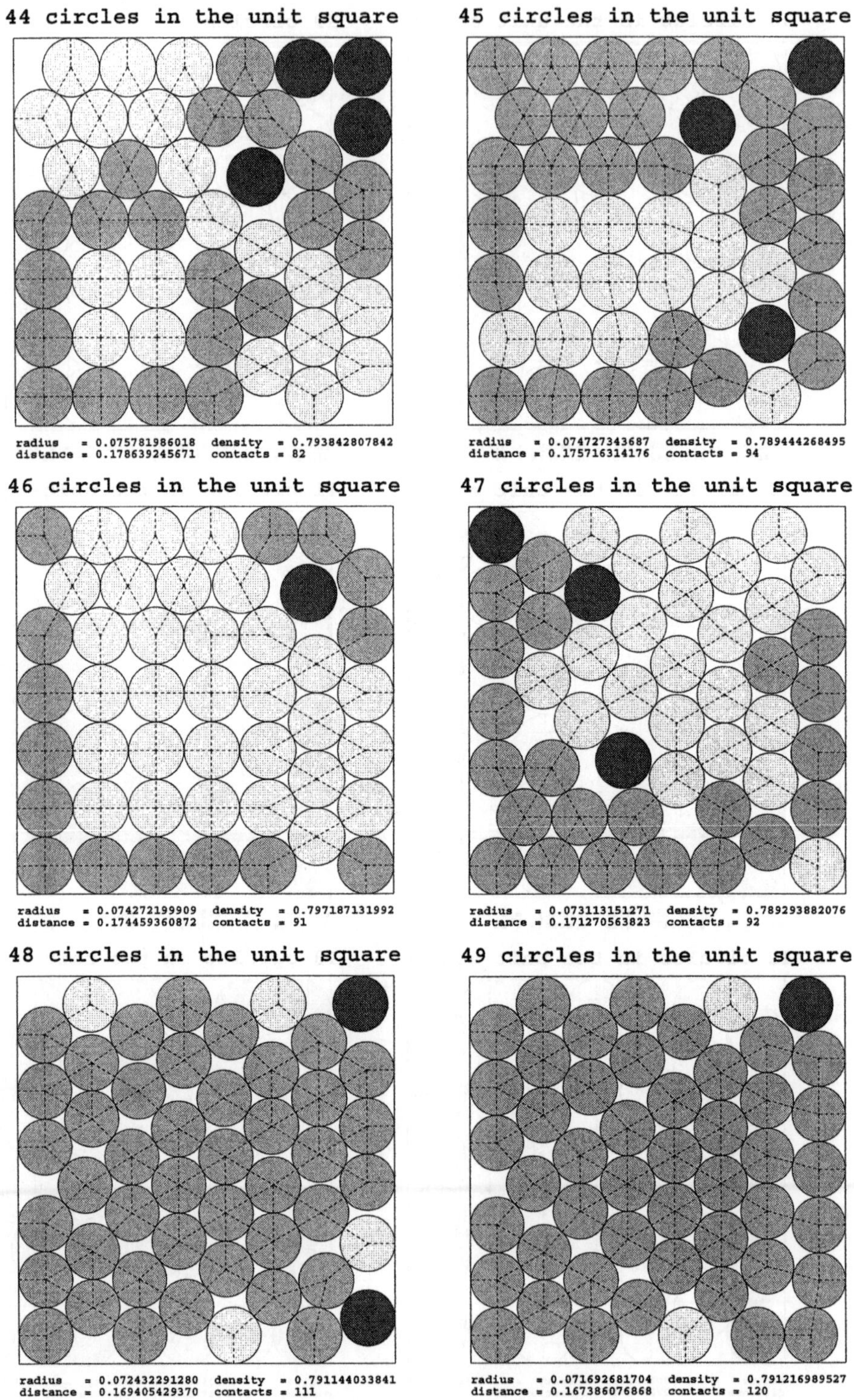

Fig. 13.8. Best known packings for 44–49 circles.

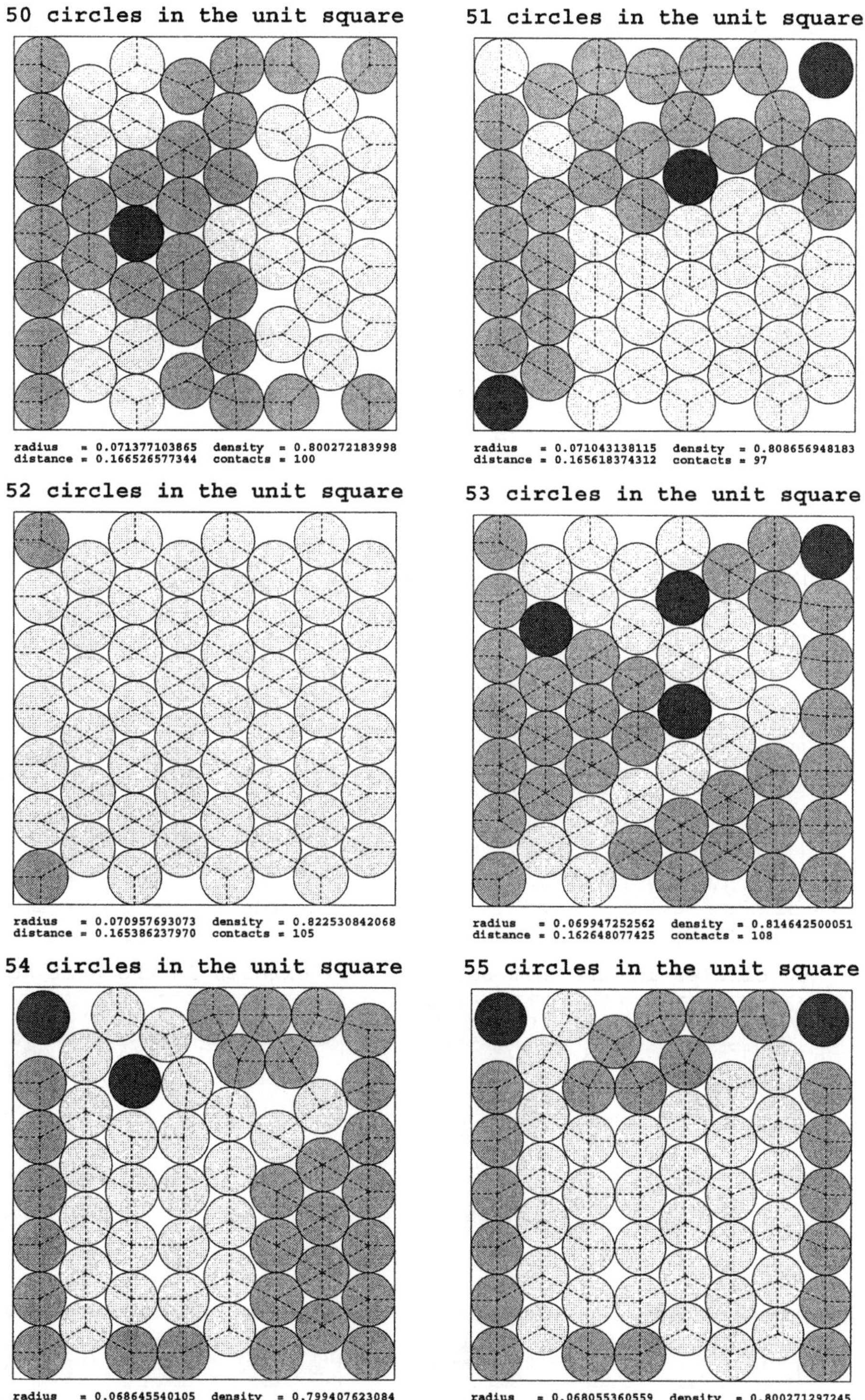

Fig. 13.9. Best known packings for 50–55 circles.

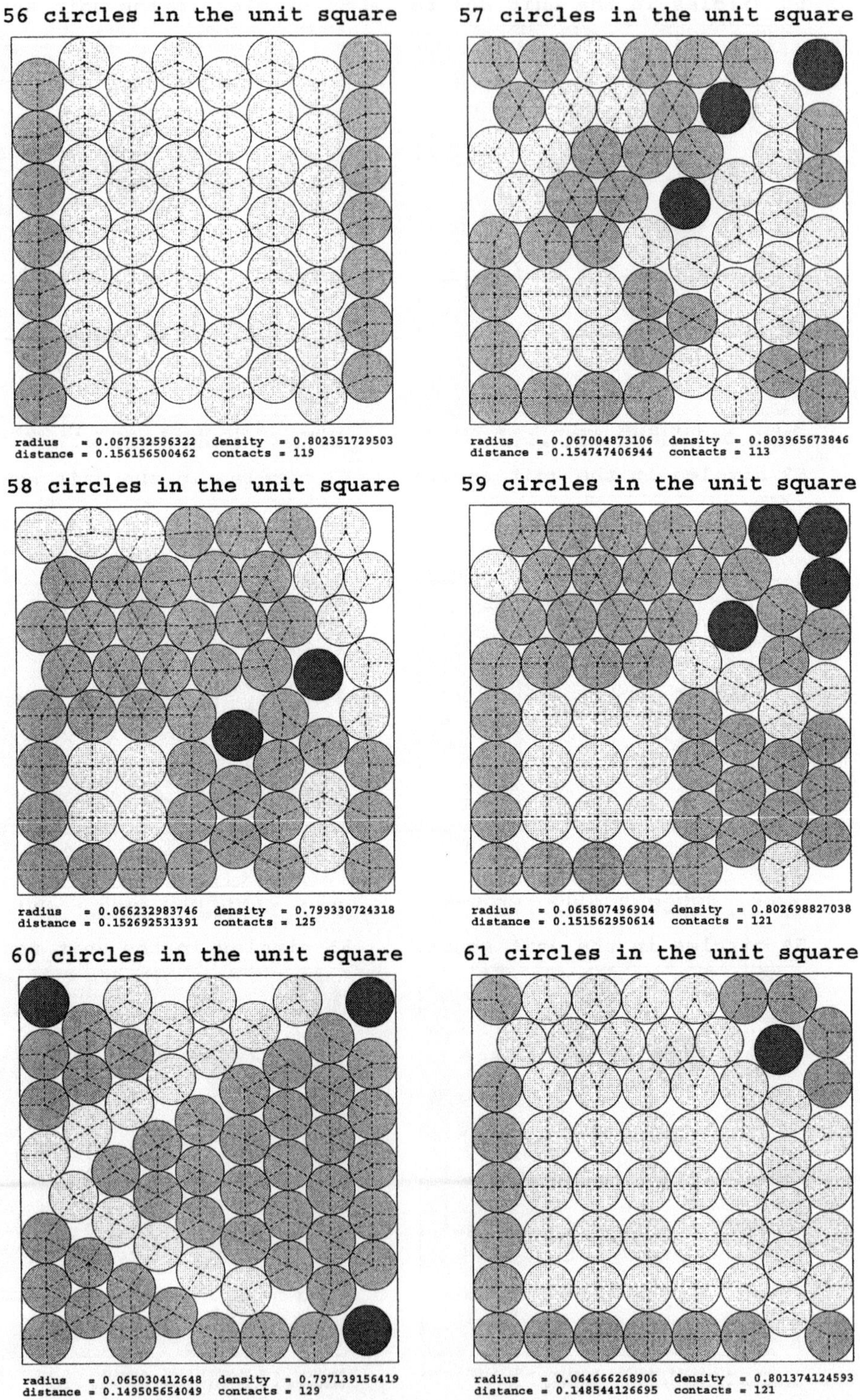

Fig. 13.10. Best known packings for 56–61 circles.

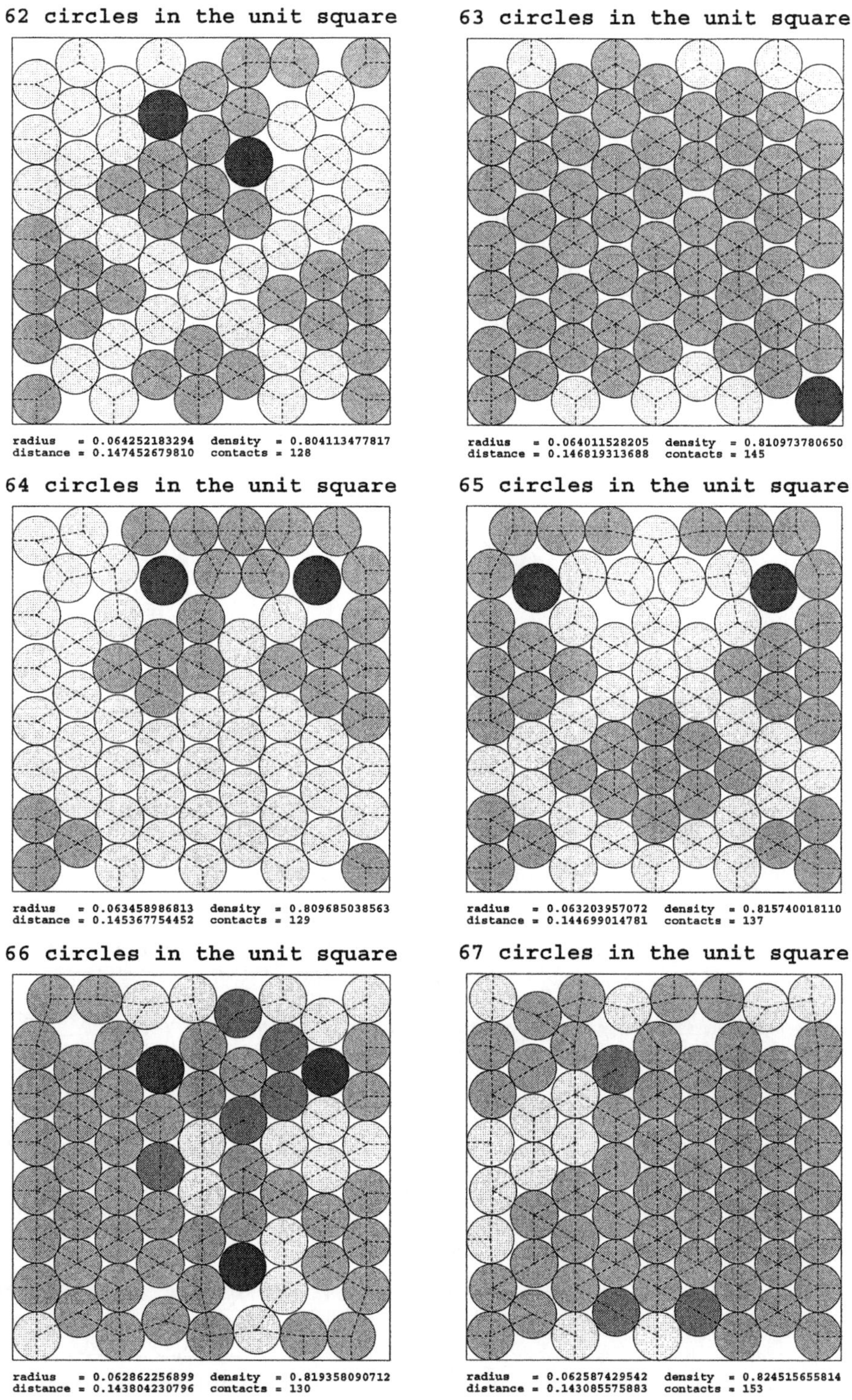

Fig. 13.11. Best known packings for 62–67 circles.

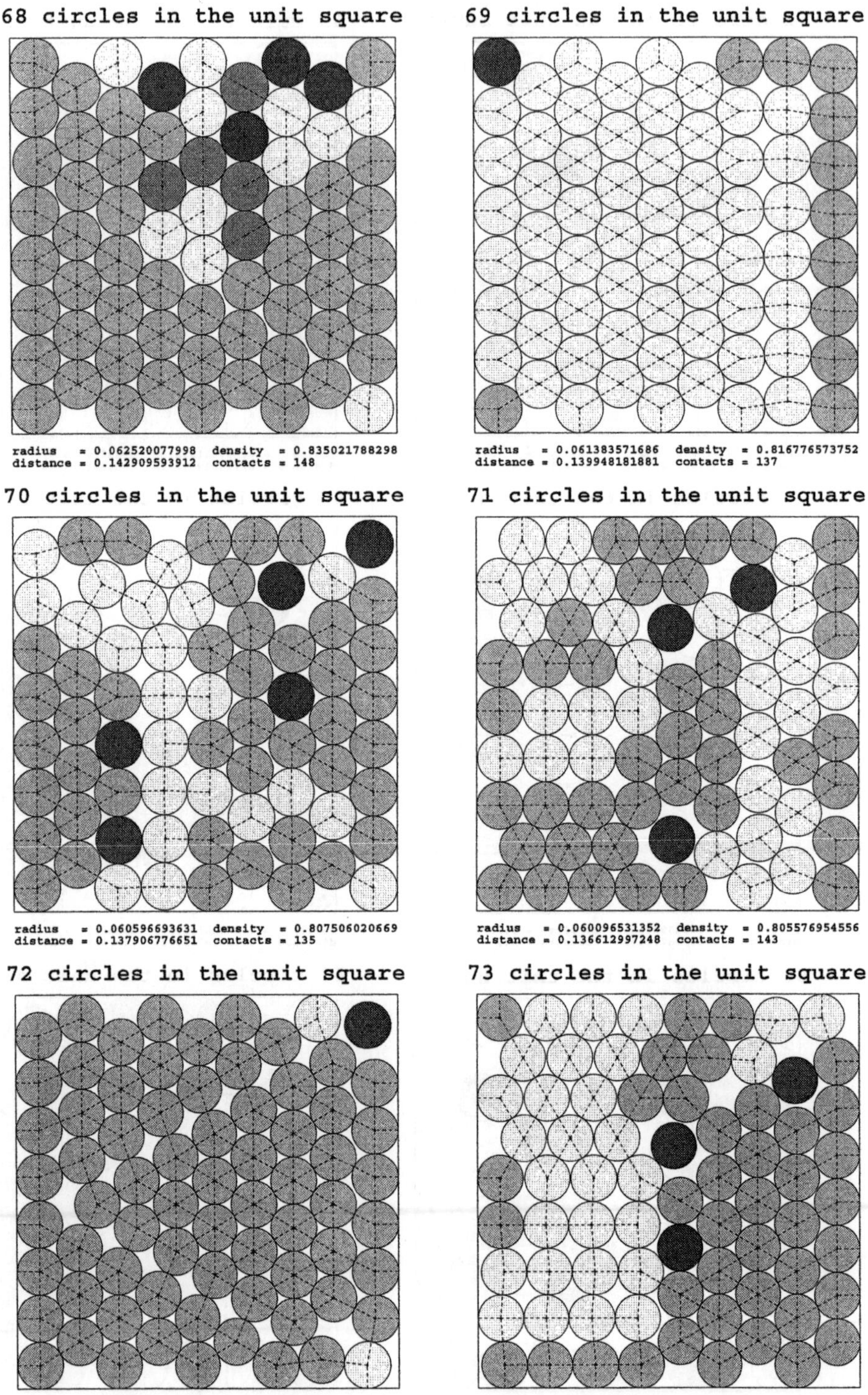

Fig. 13.12. Best known packings for 68–73 circles.

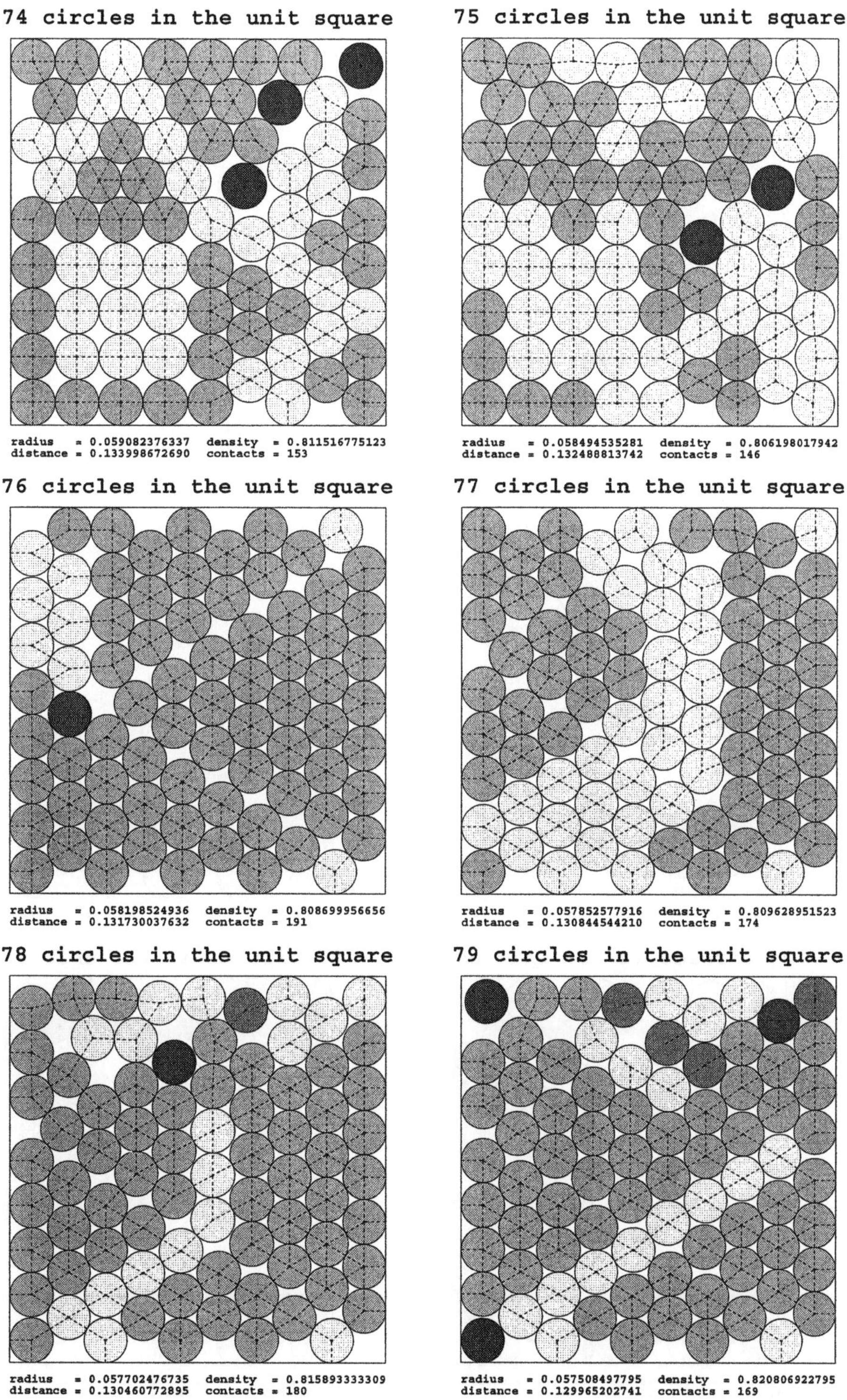

Fig. 13.13. Best known packings for 74–79 circles.

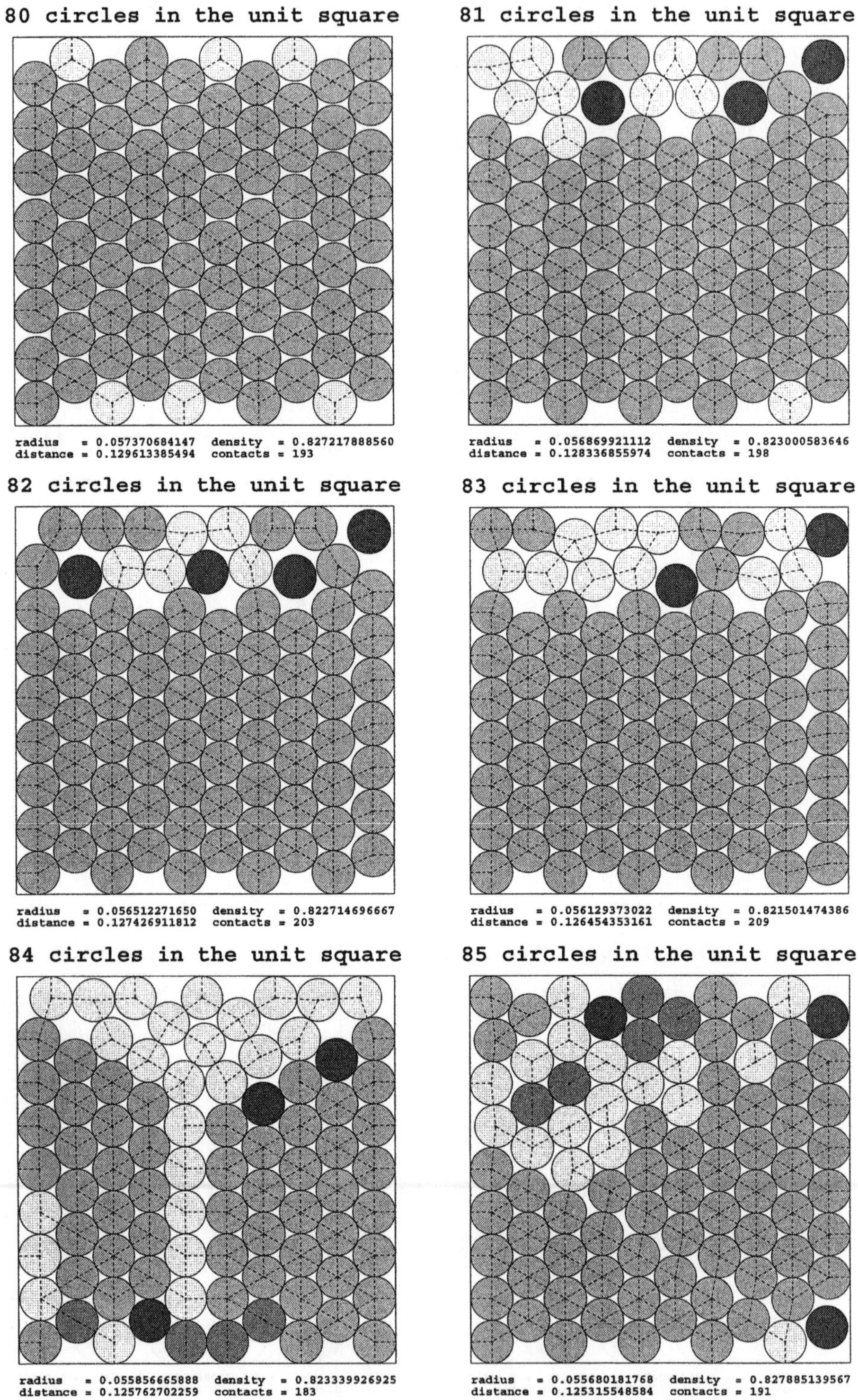

Fig. 13.14. Best known packings for 80–85 circles.

86 circles in the unit square

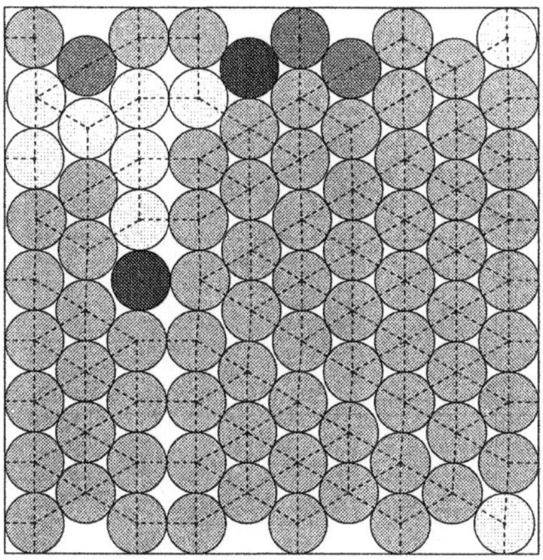

radius = 0.055572999407 density = 0.834403272277
distance = 0.125044156483 contacts = 205

87 circles in the unit square

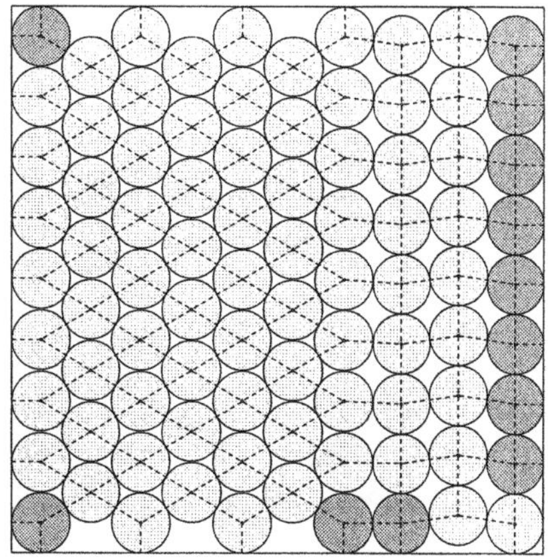

radius = 0.054695259721 density = 0.817651999581
distance = 0.122826583176 contacts = 175

88 circles in the unit square

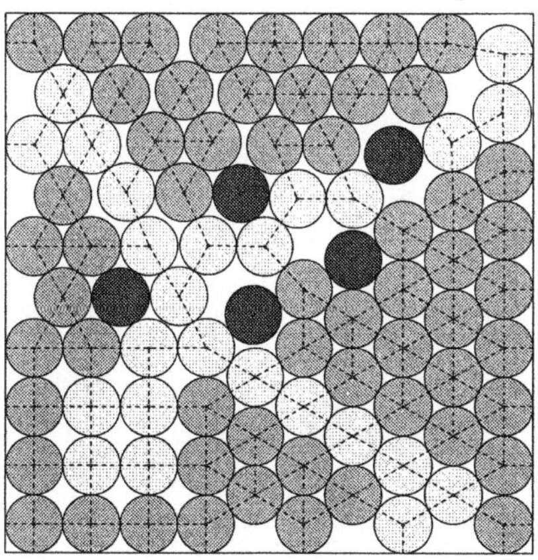

radius = 0.054406636913 density = 0.818344762893
distance = 0.122099298194 contacts = 187

89 circles in the unit square

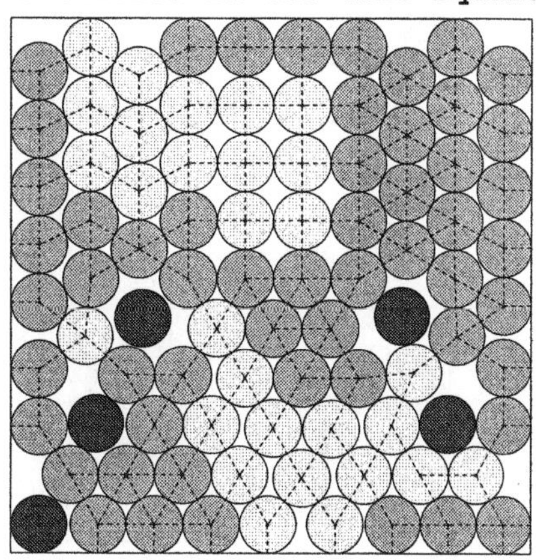

radius = 0.053947040858 density = 0.813720269423
distance = 0.120943129628 contacts = 182

90 circles in the unit square

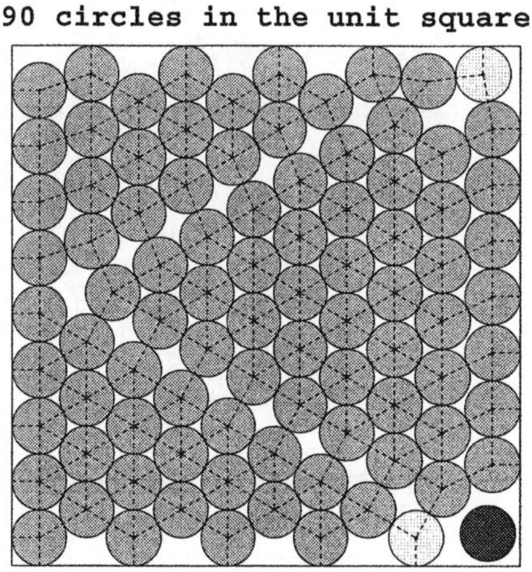

radius = 0.053749948307 density = 0.816861606108
distance = 0.120448049480 contacts = 230

91 circles in the unit square

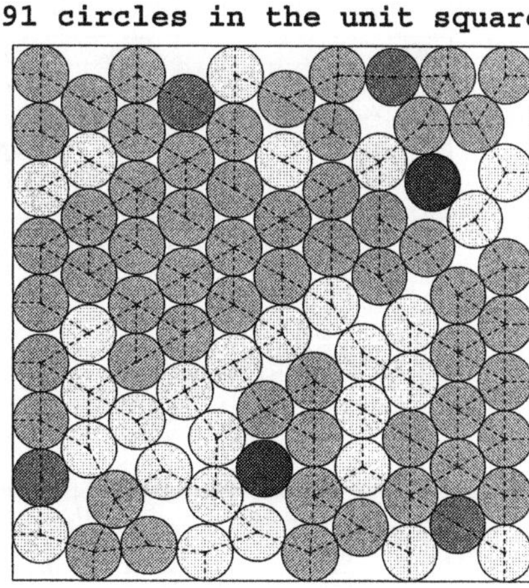

radius = 0.053496719884 density = 0.818173810461
distance = 0.119812602205 contacts = 178

Fig. 13.15. Best known packings for 86–91 circles.

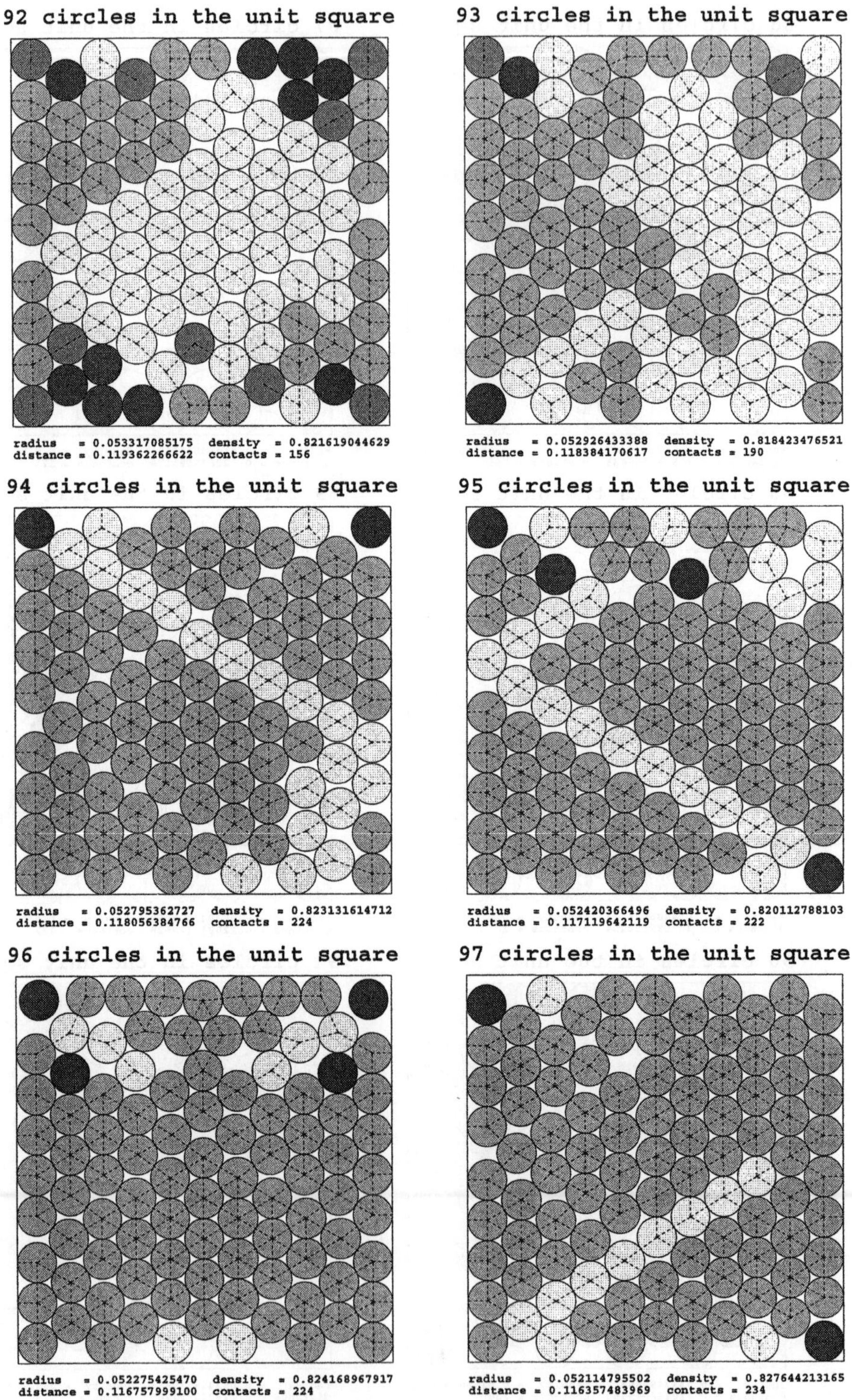

Fig. 13.16. Best known packings for 92–97 circles.

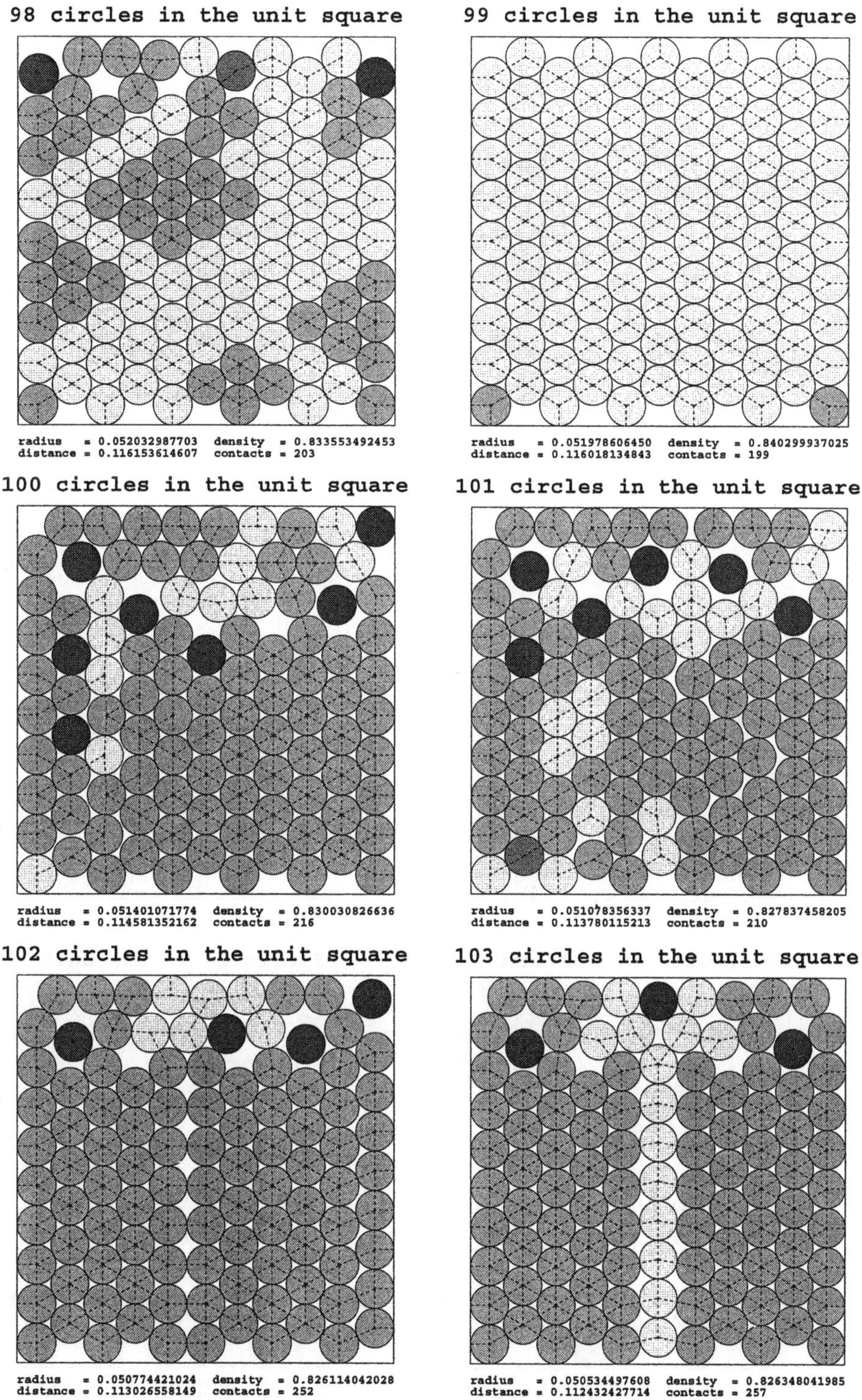

Fig. 13.17. Best known packings for 98–103 circles.

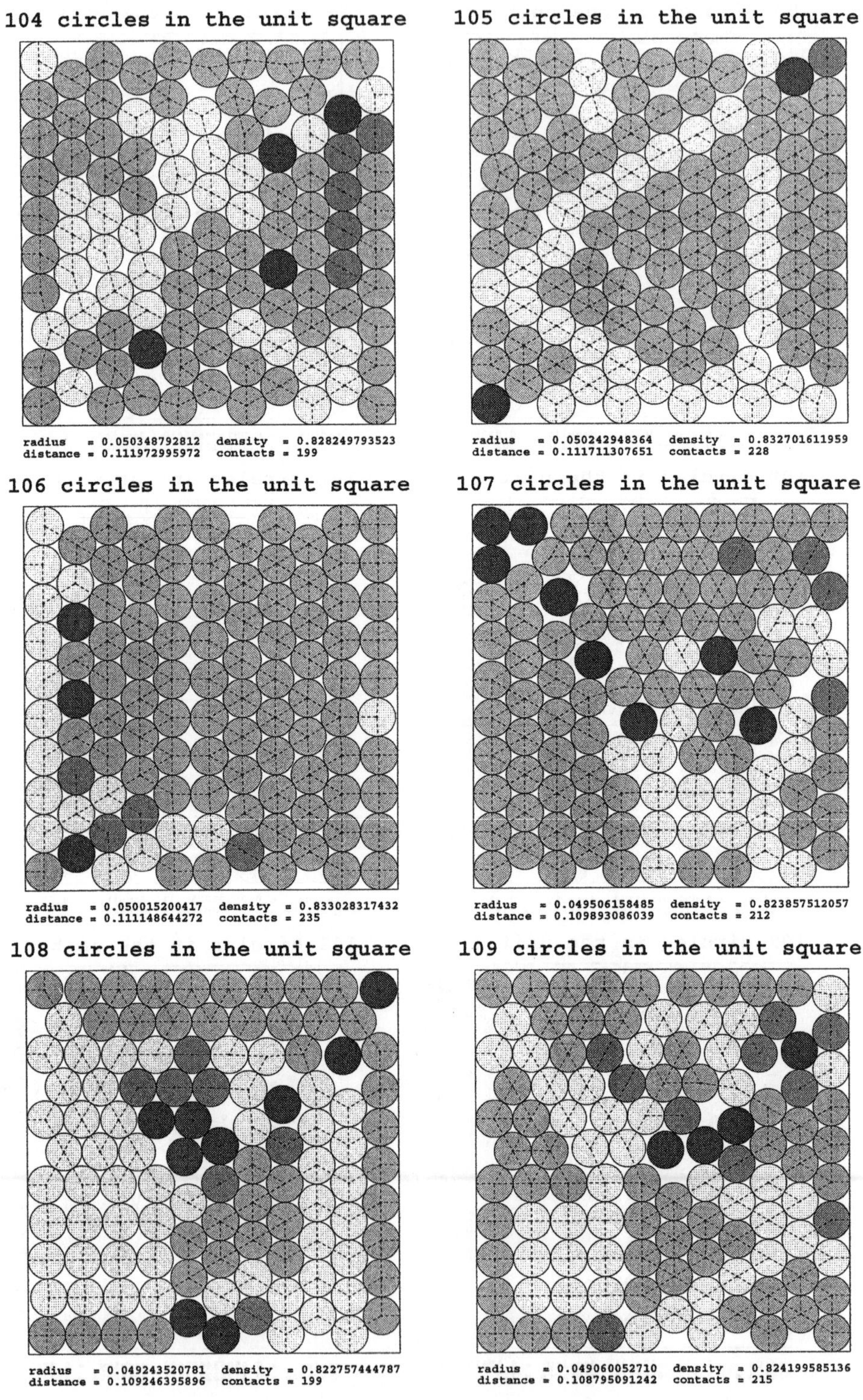

Fig. 13.18. Best known packings for 104–109 circles.

110 circles in the unit square

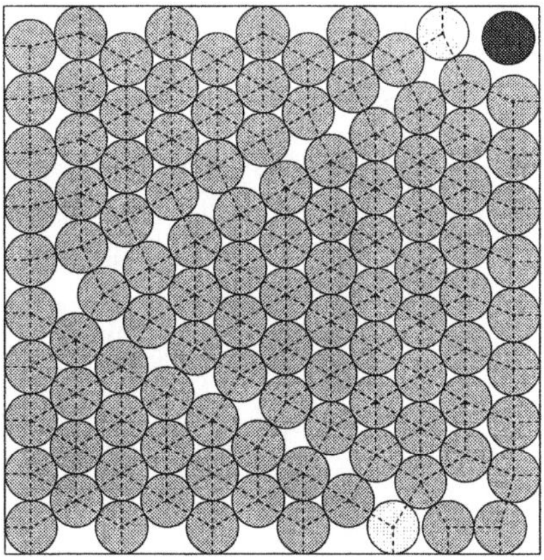

radius = 0.048779469321 density = 0.822274269044
distance = 0.108105606913 contacts = 285

111 circles in the unit square

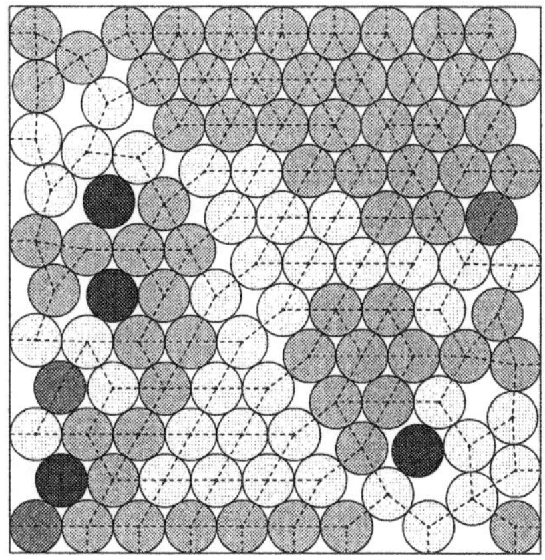

radius = 0.048643936322 density = 0.825145003148
distance = 0.107772865452 contacts = 221

112 circles in the unit square

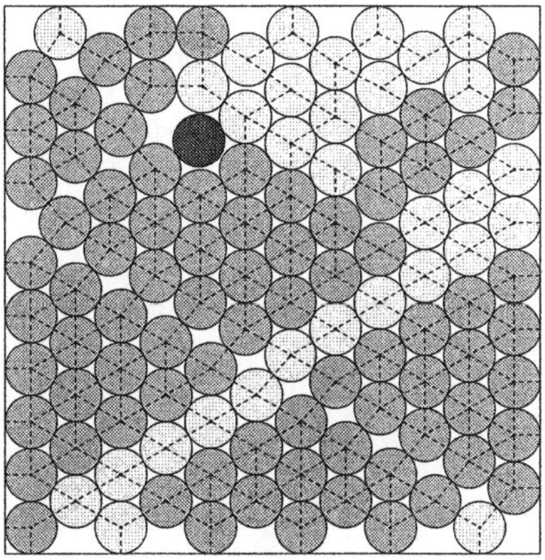

radius = 0.048445405903 density = 0.825796605619
distance = 0.107285822215 contacts = 261

113 circles in the unit square

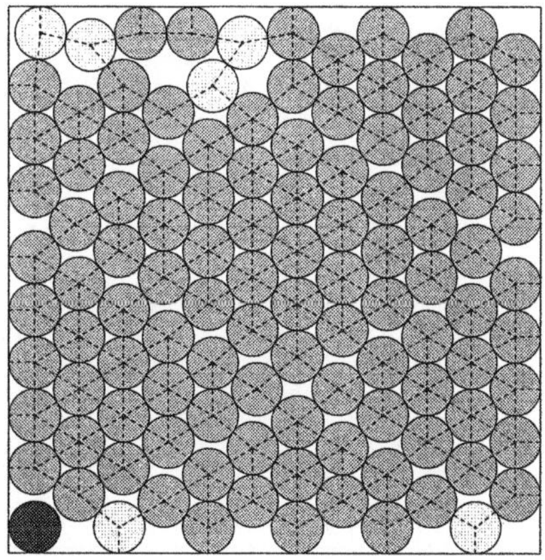

radius = 0.048257588494 density = 0.826722100560
distance = 0.106825454649 contacts = 287

114 circles in the unit square

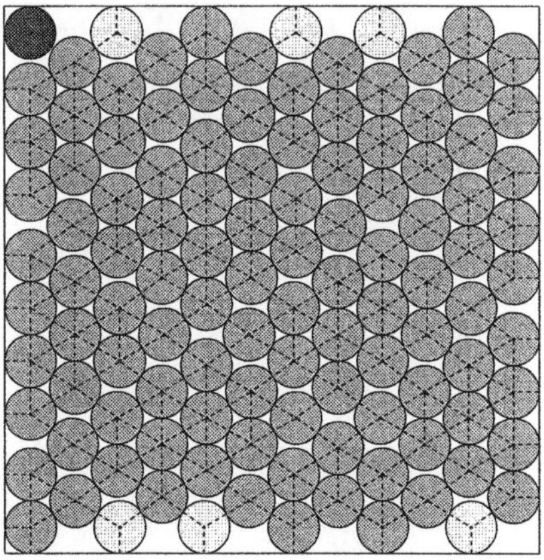

radius = 0.048167755522 density = 0.830935940404
distance = 0.106605396386 contacts = 280

115 circles in the unit square

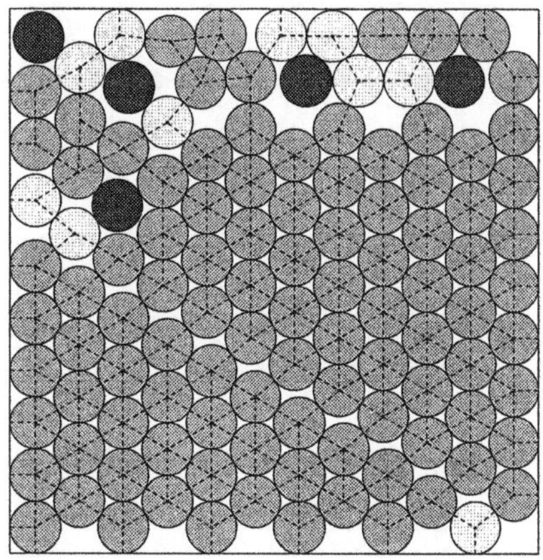

radius = 0.047914247537 density = 0.829424877813
distance = 0.105984865206 contacts = 282

Fig. 13.19. Best known packings for 110–115 circles.

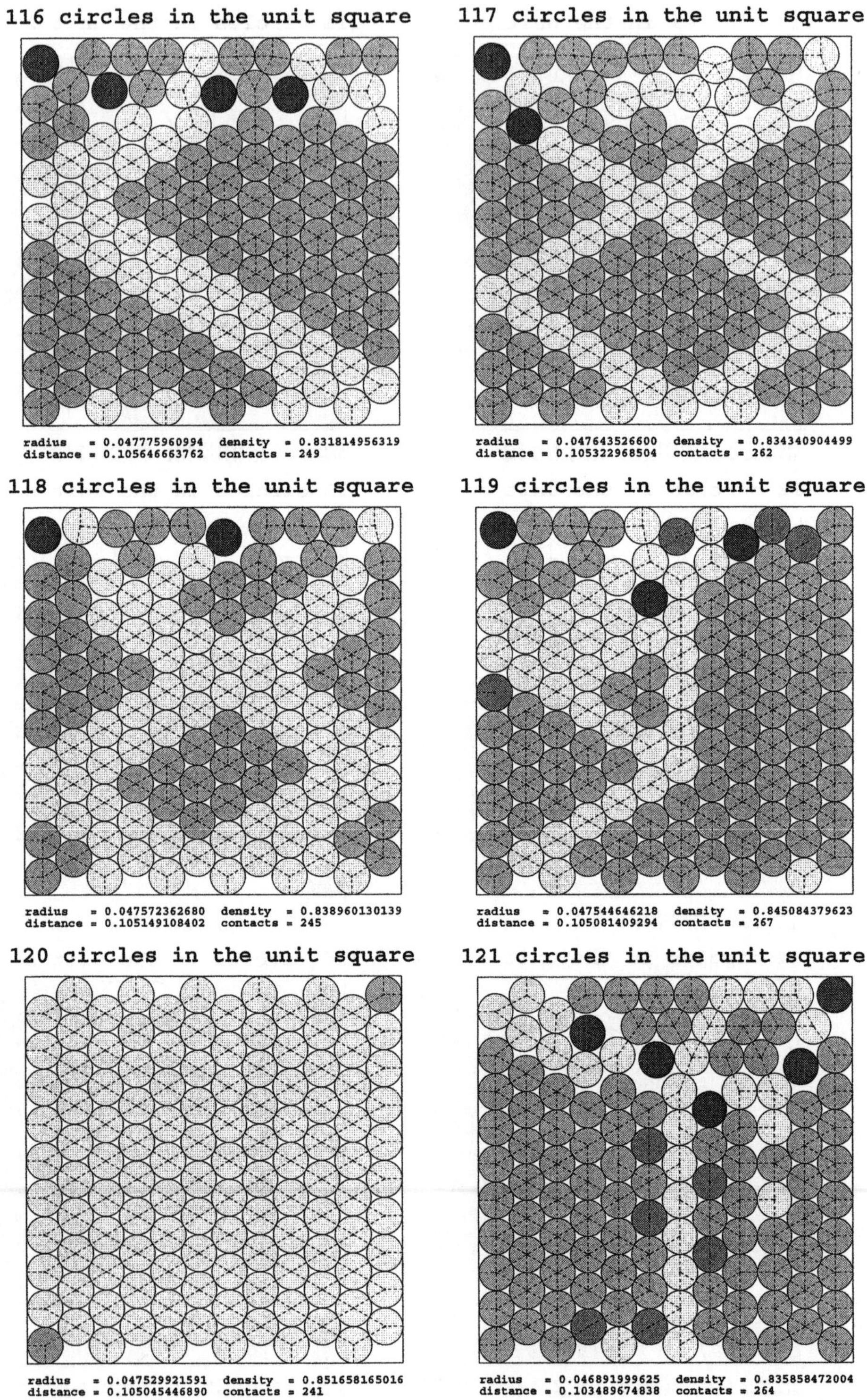

Fig. 13.20. Best known packings for 116–121 circles.

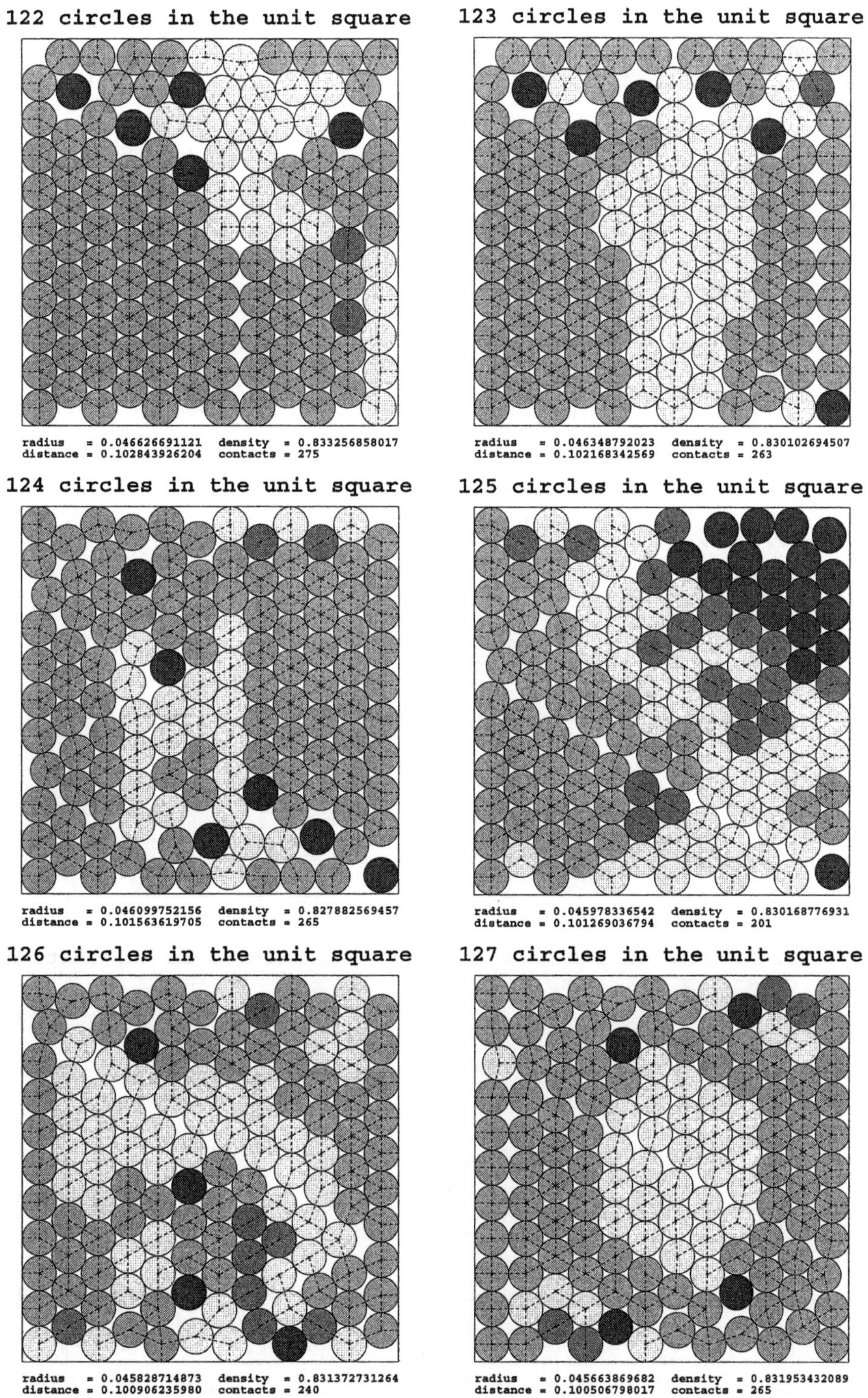

Fig. 13.21. Best known packings for 122–127 circles.

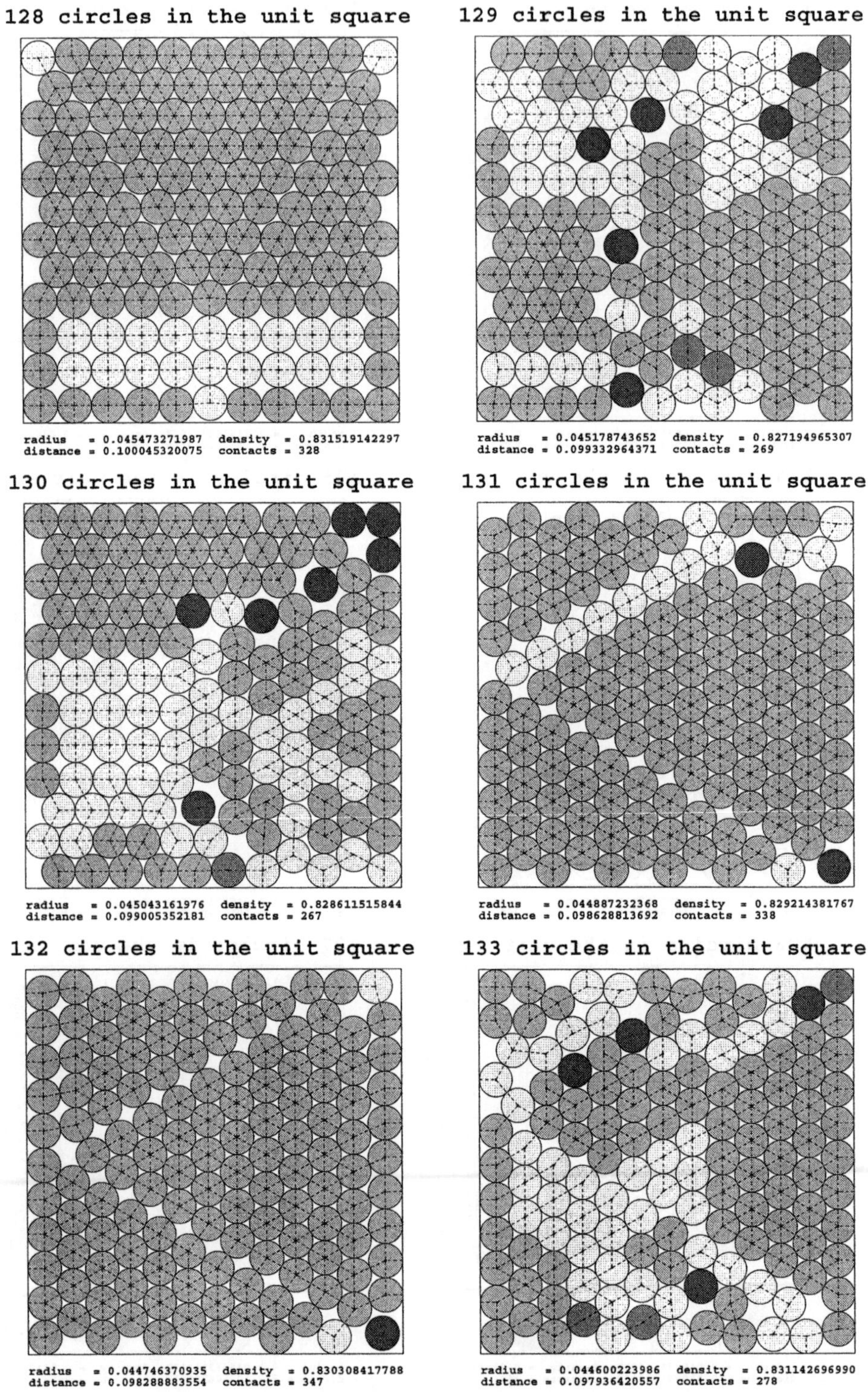

Fig. 13.22. Best known packings for 128–133 circles.

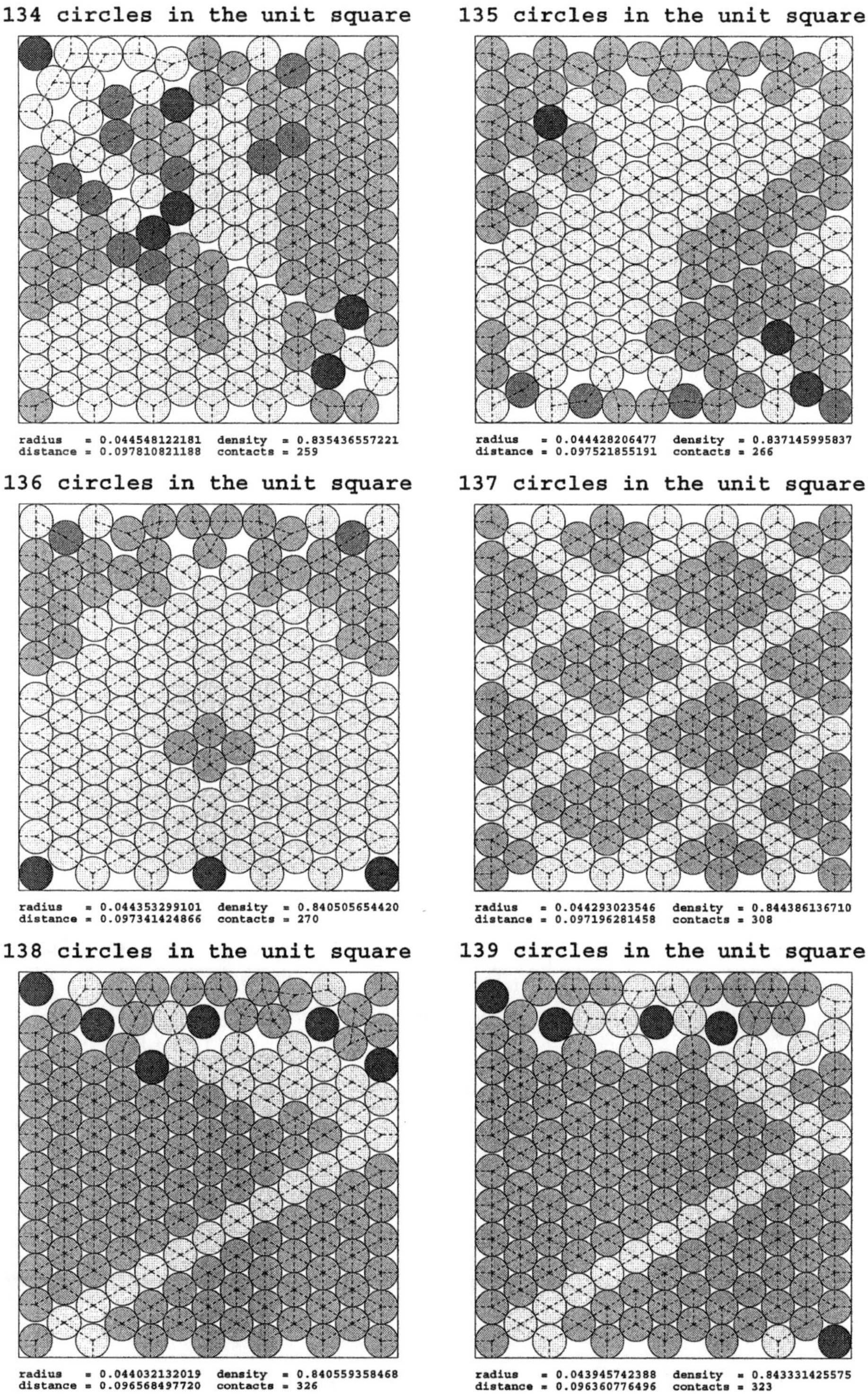

Fig. 13.23. Best known packings for 134–139 circles.

140 circles in the unit square

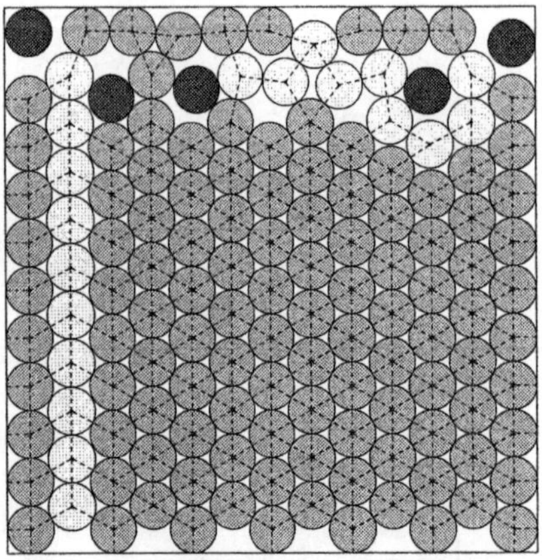

radius = 0.043804395407 density = 0.843943338354
distance = 0.096021081672 contacts = 351

141 circles in the unit square

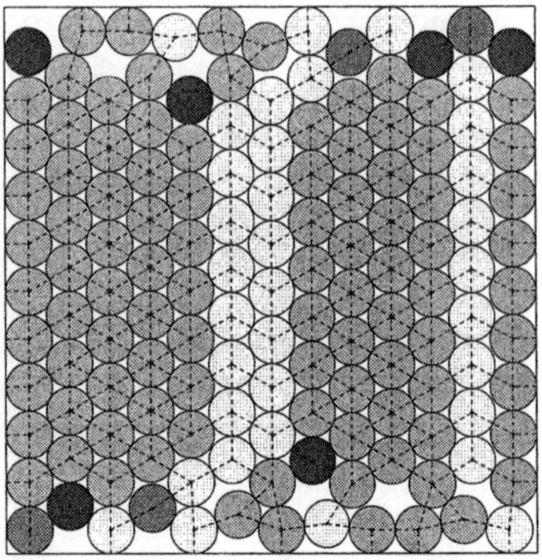

radius = 0.043658469169 density = 0.844317892156
distance = 0.095670602431 contacts = 313

142 circles in the unit square

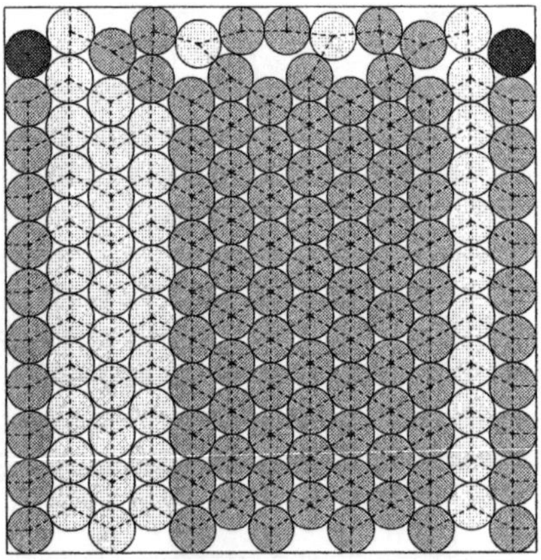

radius = 0.043576632733 density = 0.847121206057
distance = 0.095474149348 contacts = 341

143 circles in the unit square

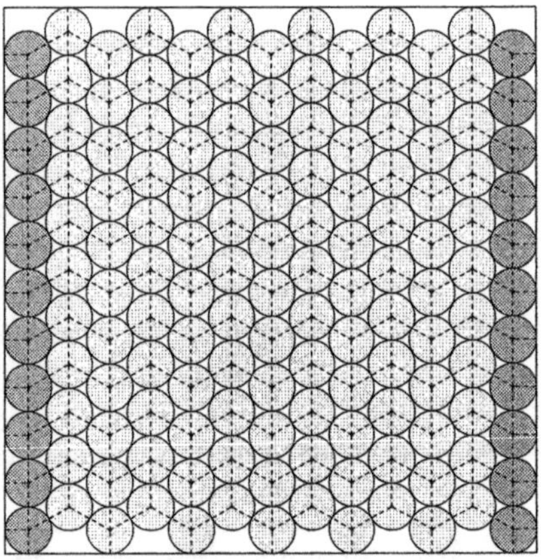

radius = 0.043530511915 density = 0.851282016761
distance = 0.095363464703 contacts = 297

144 circles in the unit square

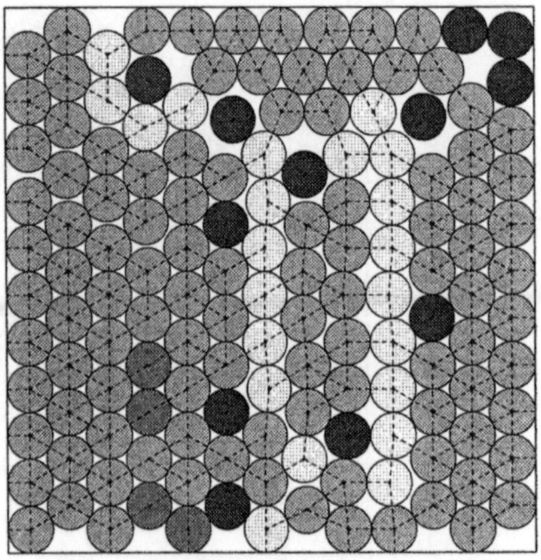

radius = 0.043002745898 density = 0.836574727482
distance = 0.094098477643 contacts = 295

145 circles in the unit square

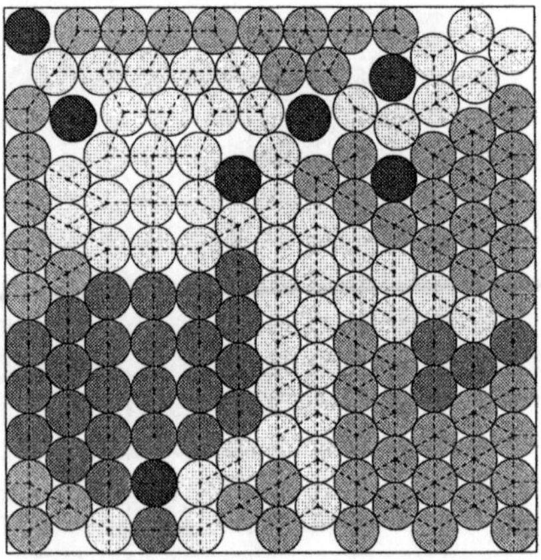

radius = 0.042871343605 density = 0.837244039299
distance = 0.093783977453 contacts = 257

Fig. 13.24. Best known packings for 140–145 circles.

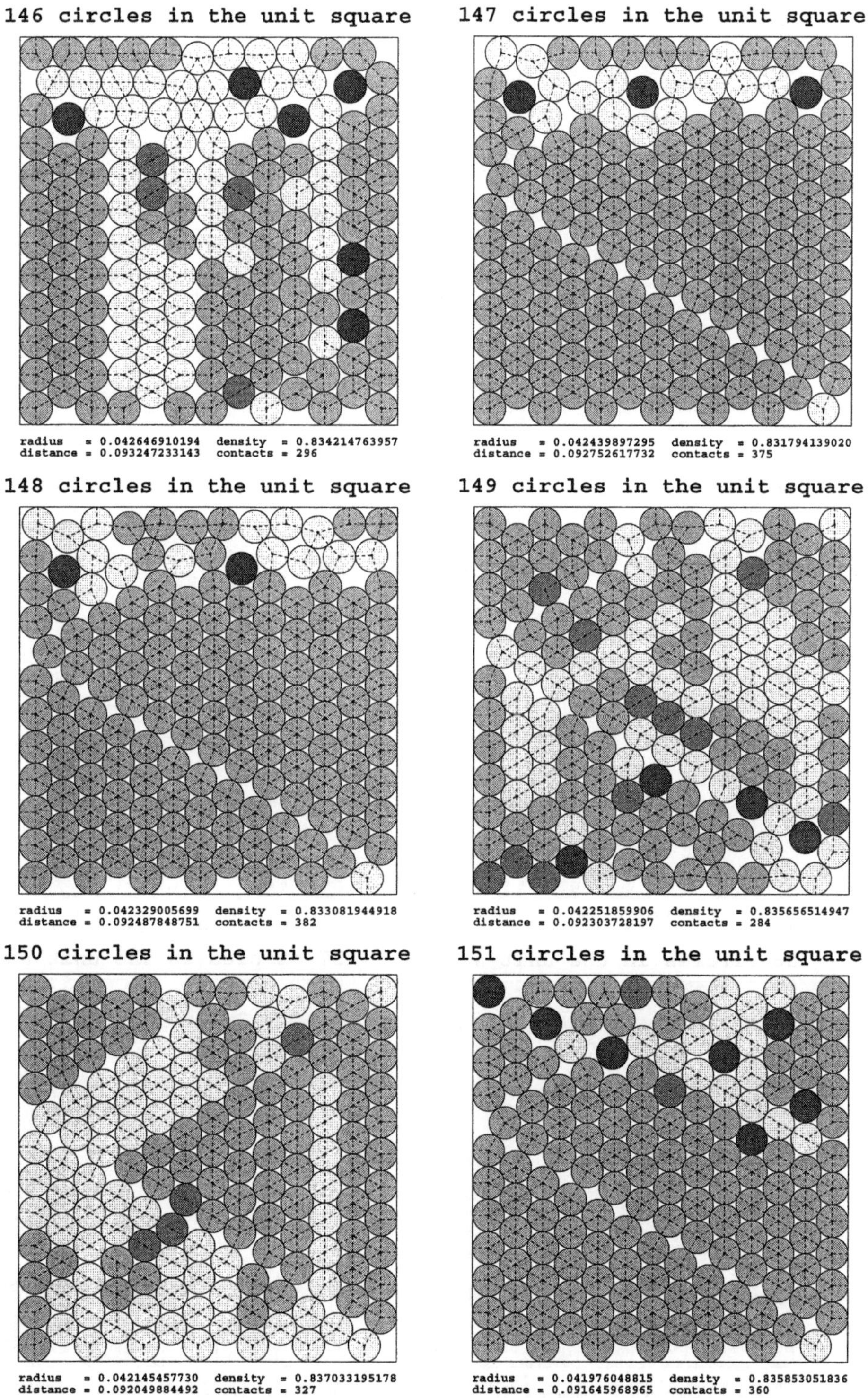

Fig. 13.25. Best known packings for 146–151 circles.

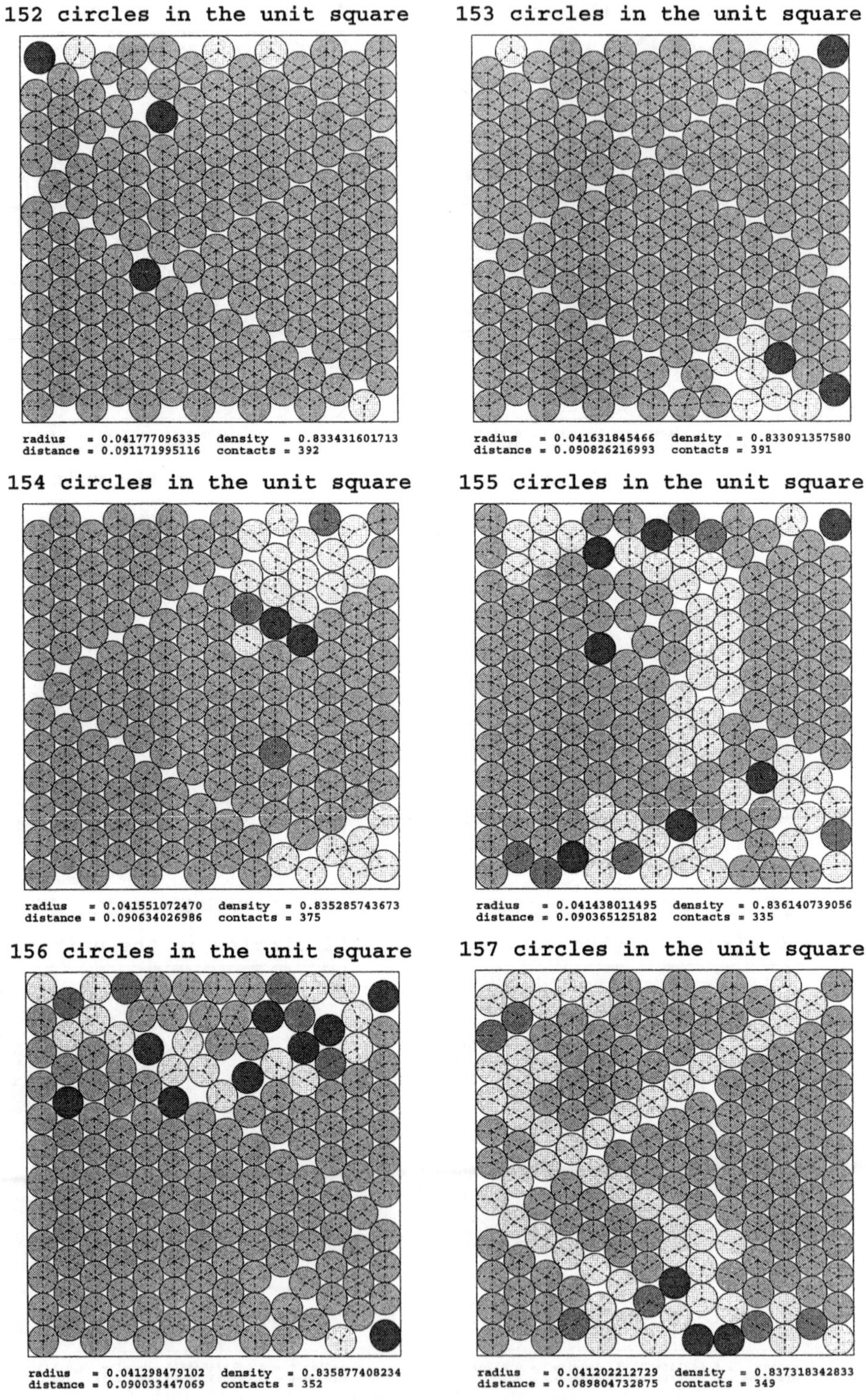

Fig. 13.26. Best known packings for 152–157 circles.

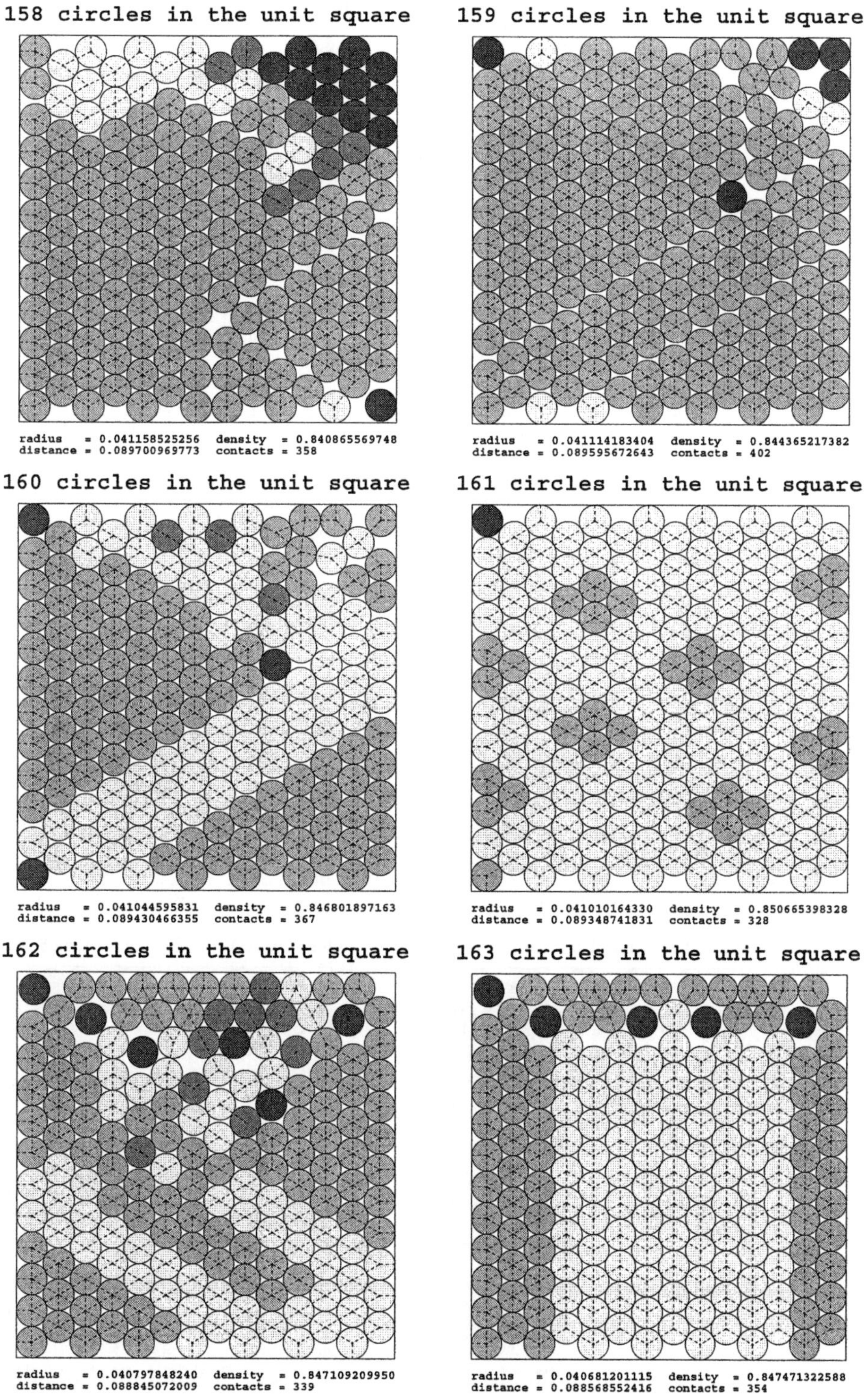

Fig. 13.27. Best known packings for 158–163 circles.

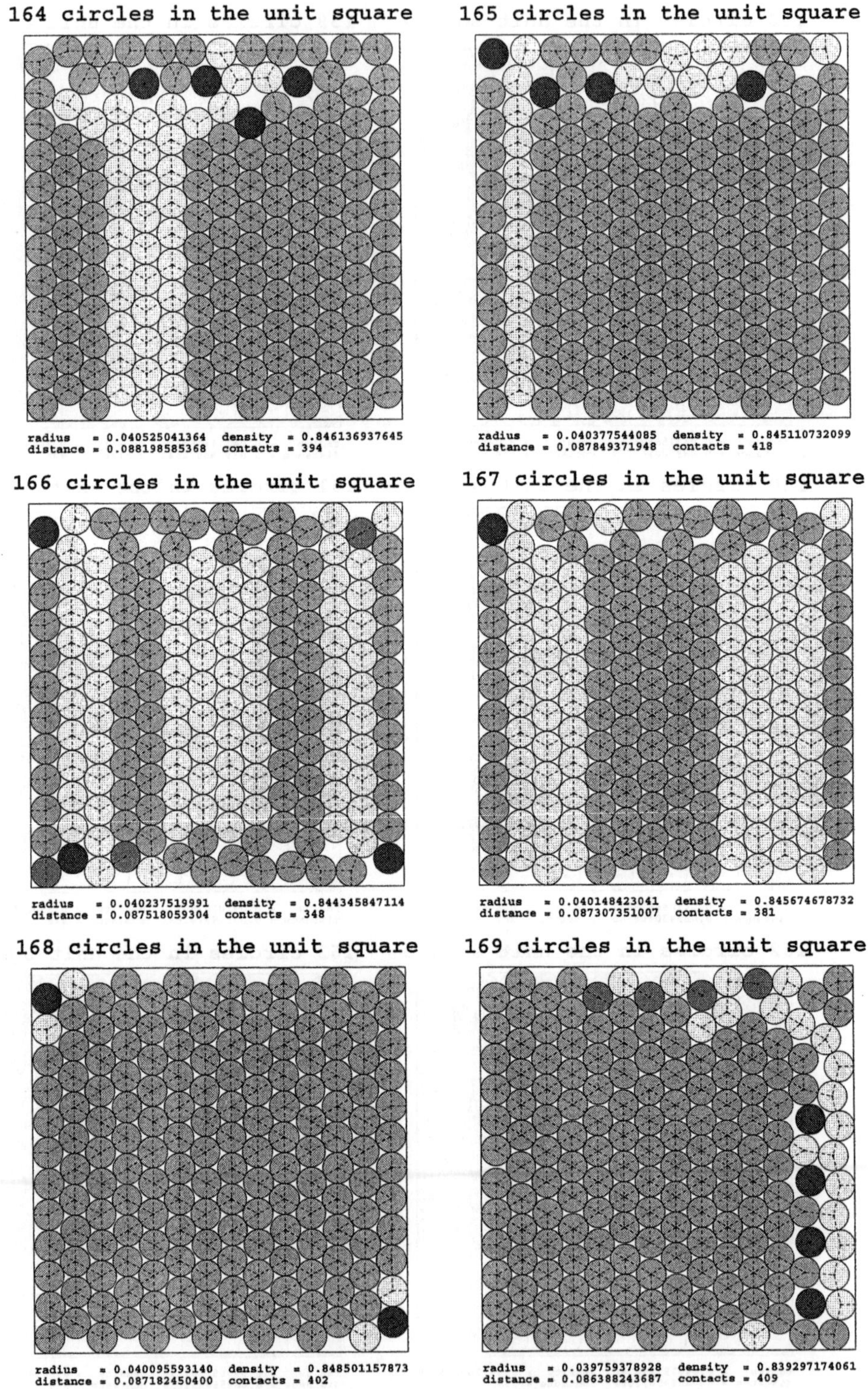

Fig. 13.28. Best known packings for 164–169 circles.

170 circles in the unit square

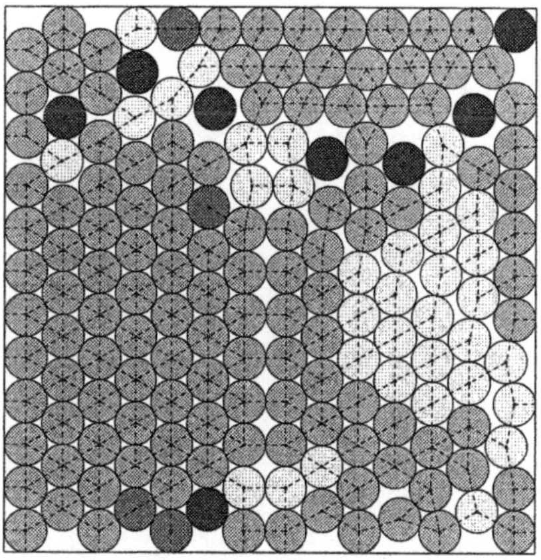

radius = 0.039585933812 density = 0.836913517436
distance = 0.085978984395 contacts = 364

171 circles in the unit square

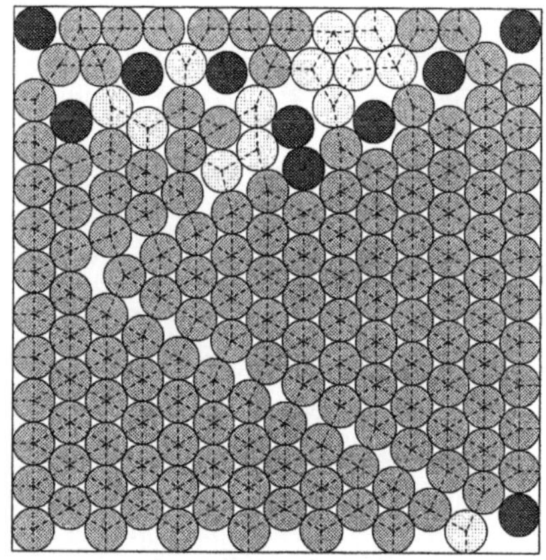

radius = 0.039434198868 density = 0.835395300146
distance = 0.085621204987 contacts = 416

172 circles in the unit square

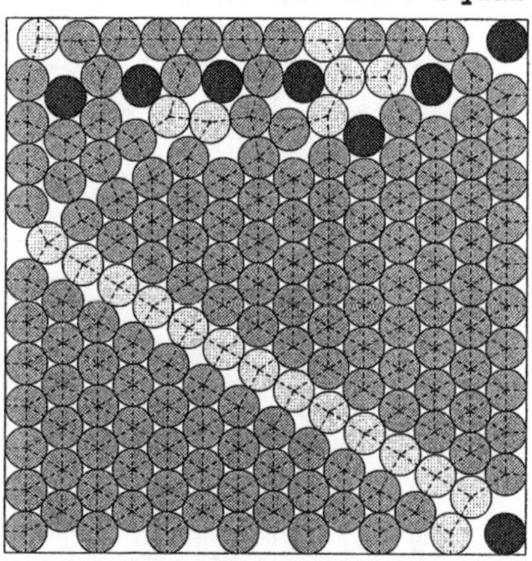

radius = 0.039318677249 density = 0.835364696522
distance = 0.085348971855 contacts = 418

173 circles in the unit square

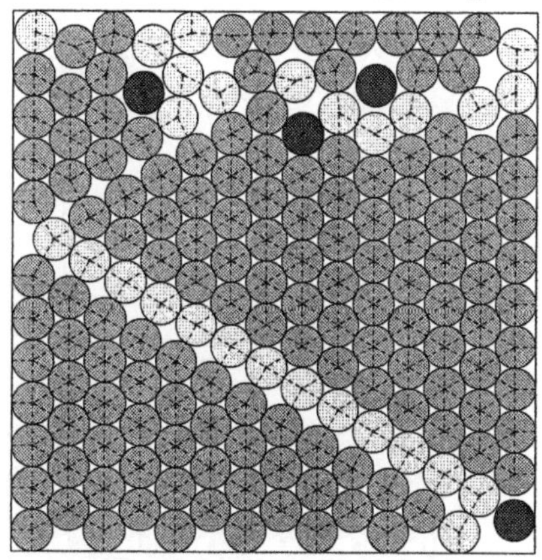

radius = 0.039236445096 density = 0.836710619012
distance = 0.085155270372 contacts = 429

174 circles in the unit square

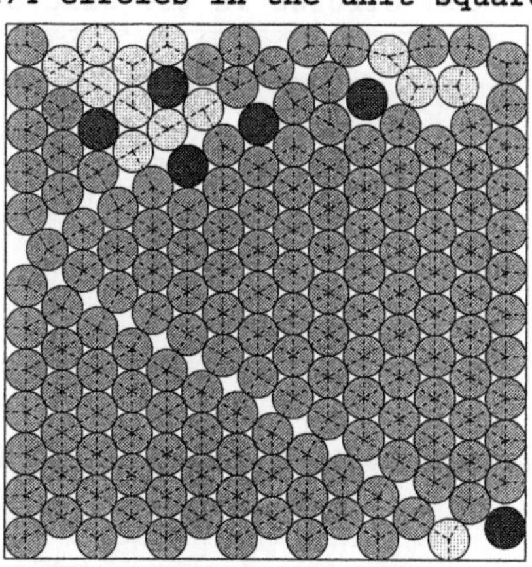

radius = 0.039179543023 density = 0.839107984106
distance = 0.085021275487 contacts = 441

175 circles in the unit square

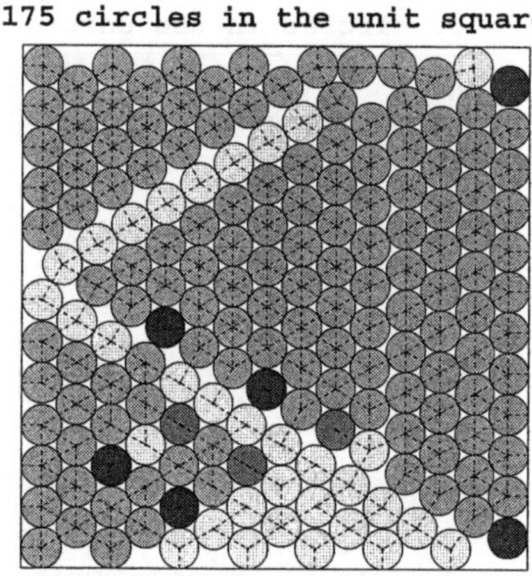

radius = 0.039046149820 density = 0.838193626231
distance = 0.084707286435 contacts = 399

Fig. 13.29. Best known packings for 170–175 circles.

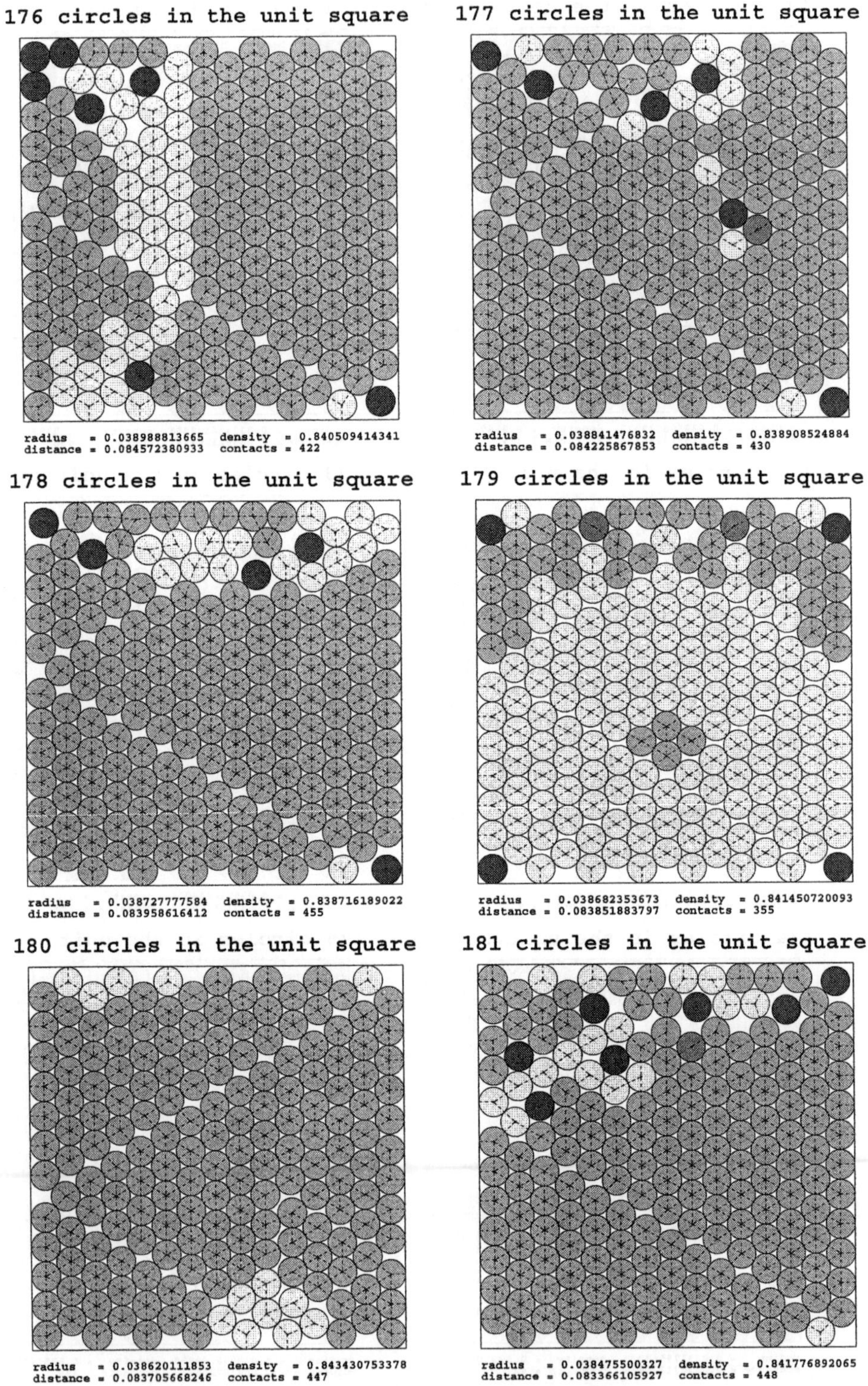

Fig. 13.30. Best known packings for 176–181 circles.

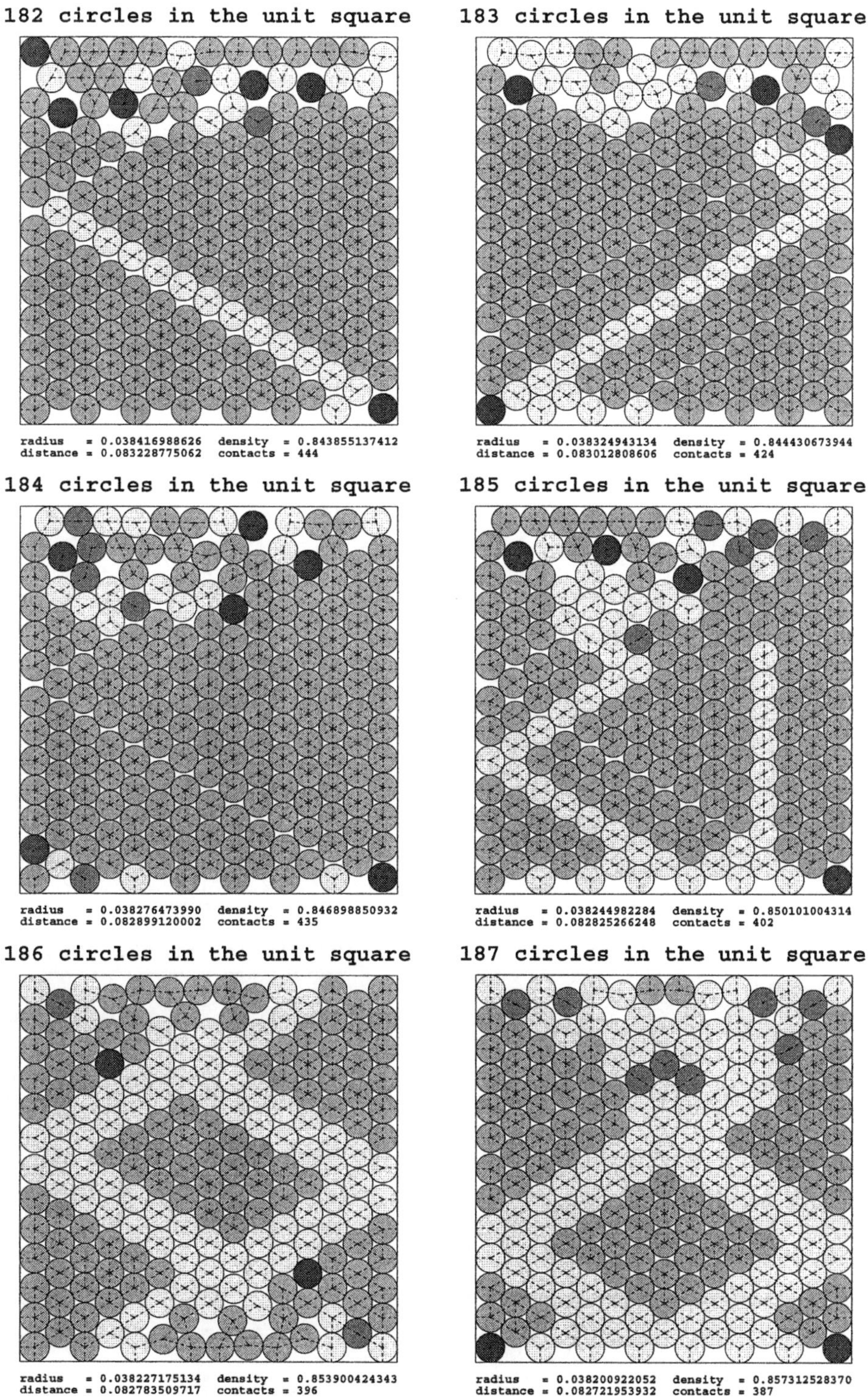

Fig. 13.31. Best known packings for 182–187 circles.

188 circles in the unit square

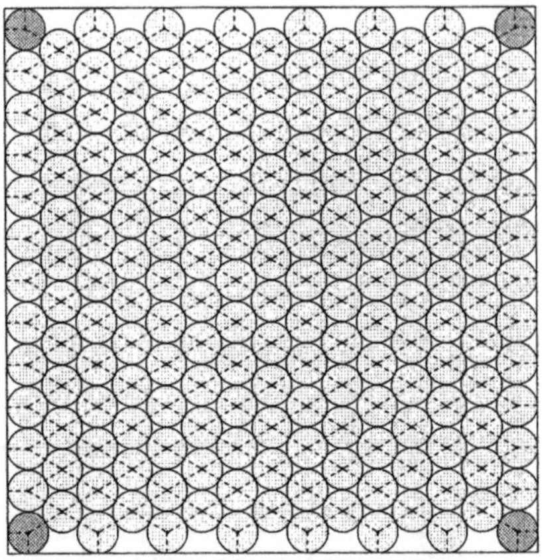

radius = 0.038188623136 density = 0.861342195310
distance = 0.082693118985 contacts = 378

189 circles in the unit square

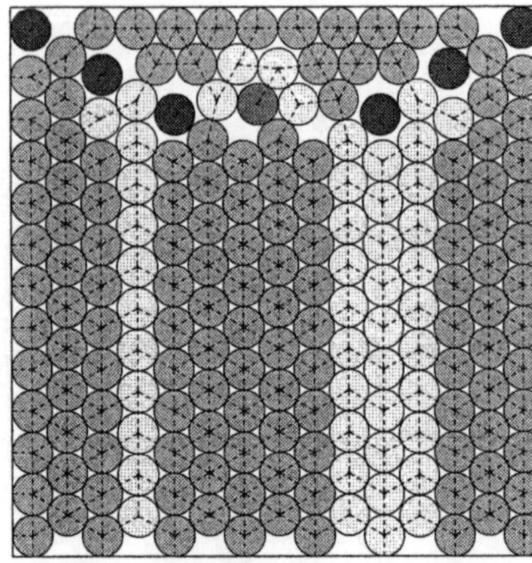

radius = 0.037866552704 density = 0.851379553528
distance = 0.081938567583 contacts = 448

190 circles in the unit square

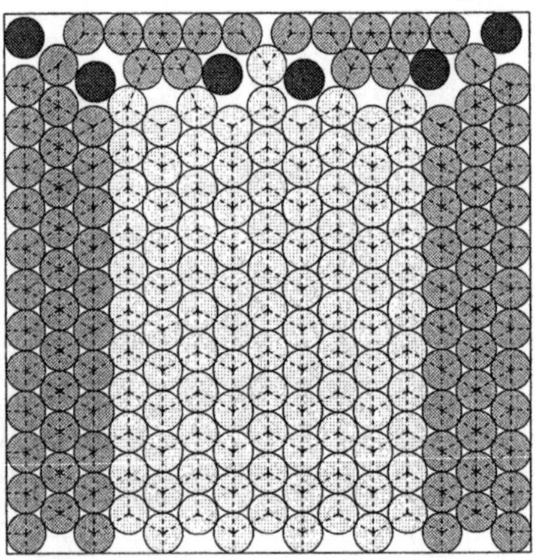

radius = 0.037695999979 density = 0.848191685066
distance = 0.081539419899 contacts = 421

191 circles in the unit square

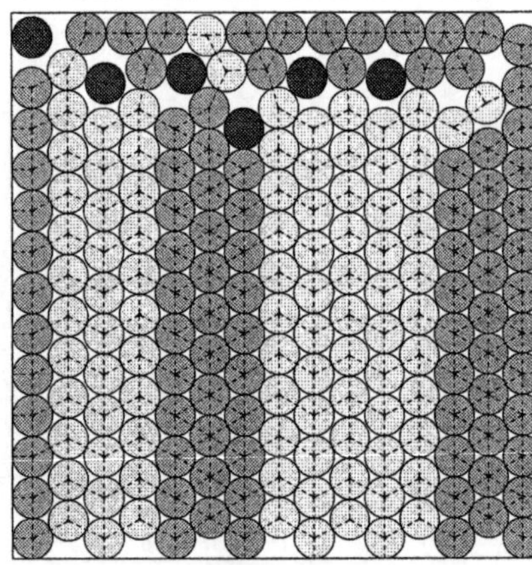

radius = 0.037544127565 density = 0.845799206937
distance = 0.081184237898 contacts = 417

192 circles in the unit square

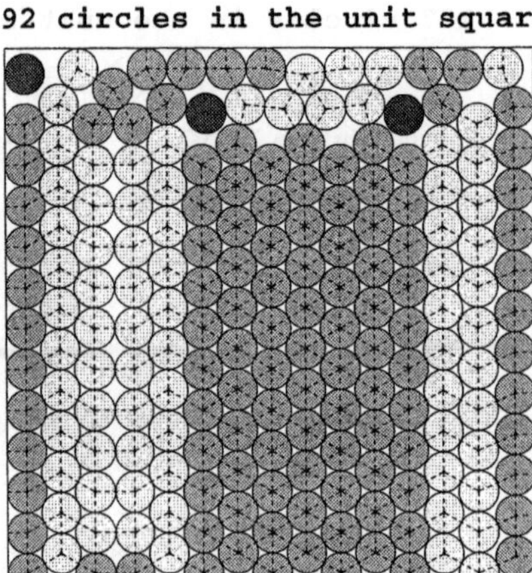

radius = 0.037431098348 density = 0.845115842743
distance = 0.080920049346 contacts = 447

193 circles in the unit square

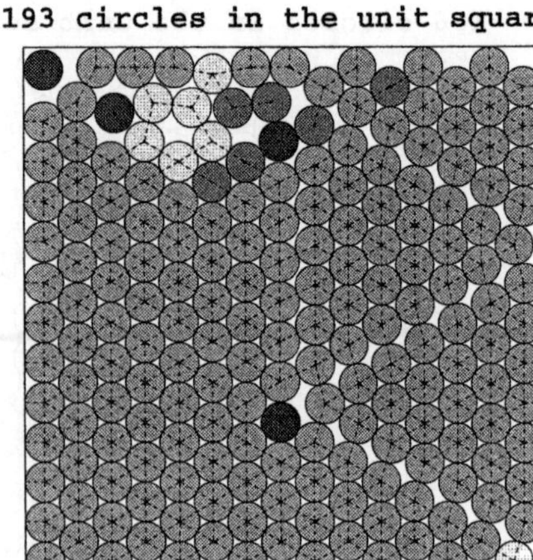

radius = 0.037310270862 density = 0.844041859326
distance = 0.080637776273 contacts = 484

Fig. 13.32. Best known packings for 188–193 circles.

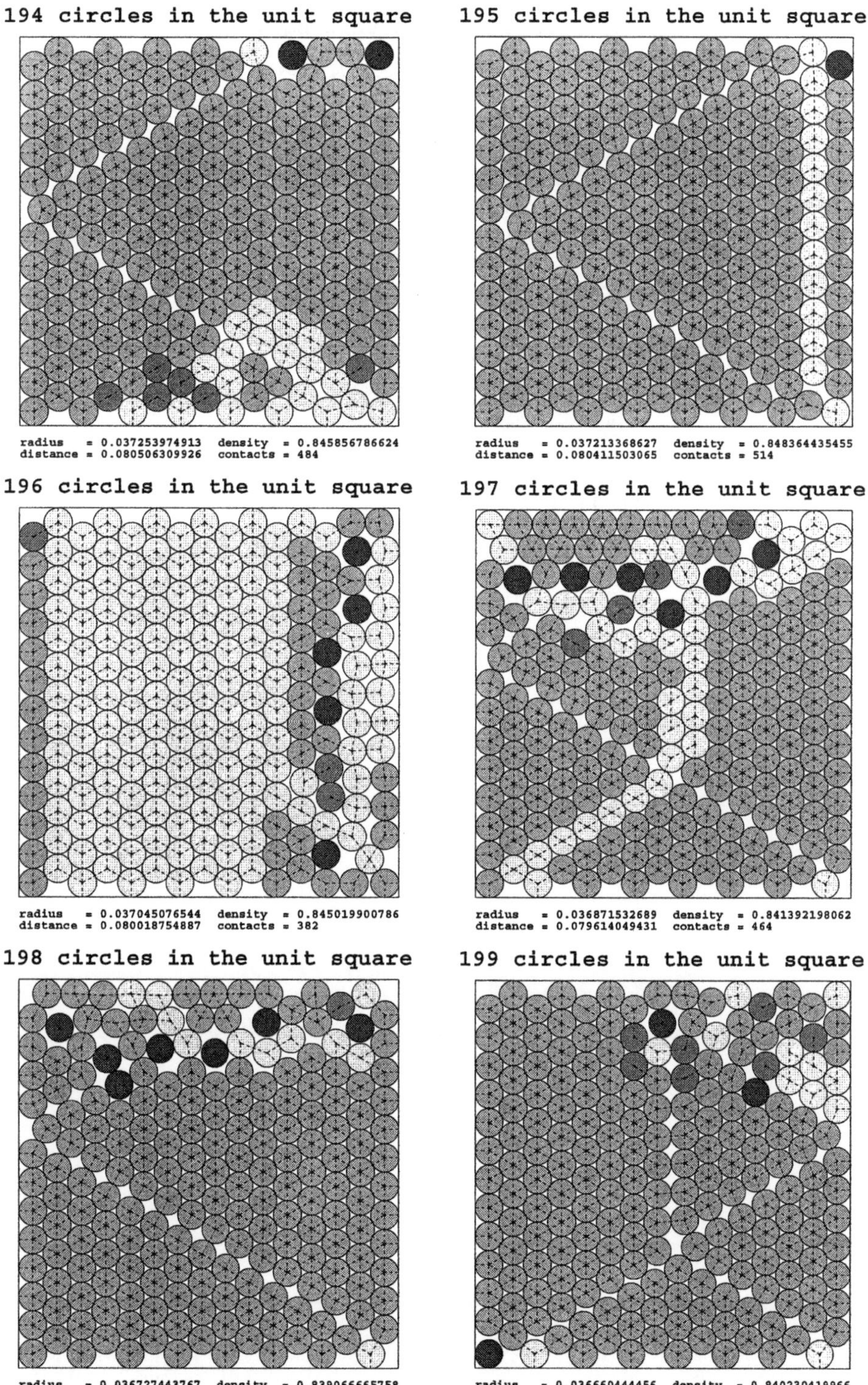

Fig. 13.33. Best known packings for 194–199 circles.

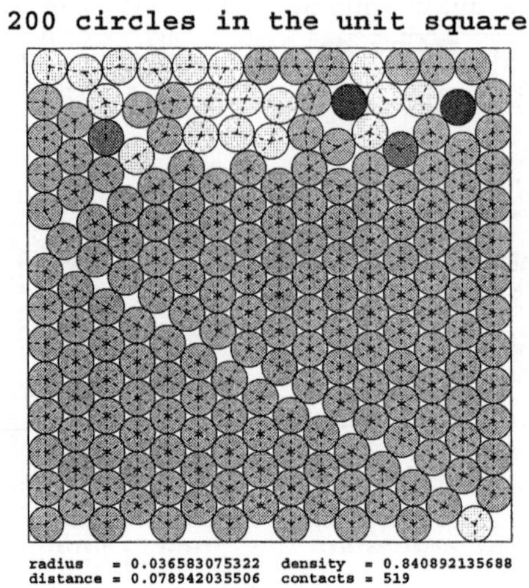

Fig. 13.34. Best known packing for 200 circles.

Bibliography

1. B. Addis, M. Locatelli, and F. Schoen: Packing circles in a square: new putative optima obtained via global optimization. *Optimization Online* (2005). Available at `http://www.optimization-online.org/DB_HTML/2005/03/1088.html`.
2. I. Althöfer and K. U. Koschnick: On the convergence of threshold accepting. *Applied Mathematics and Optimization* 24 (1991) 183–195.
3. P. Ament and G. Blind: Packing equal circles in a square. *Studia Scientiarum Mathematicarum Hungarica* 36 (2000) 313–316.
4. D. Baer and W. Mögling: Über Punktverteilungen im Quadrat mit speziellem Abstand. In German. *Wissenschaftliche Zeitschrift Pädagogische Hochschule Erfurt Mühlhausen* 23 (1987) 116–120.
5. D. W. Boll, J. Donovan, R. L. Graham, and B. D. Lubachevsky: Improving Dense Packings of Equal Disks in a Square. *Electronic J. of Combinatorics* 7 (2000) #R46. Available at `http://www.combinatorics.org/Volume_7/PostScriptfiles/v7i1r46.ps`.
6. Wolfgangi Bolyai de Bolya: *Tentamen iuventutem studiosam in elementa matheseos purae elementaris ac sublimioris methodo intuitiva evidentiaque huic propria introducendi, cum appendice triplici.* In Latin. Marosvásárhelyini, 1832–33. Second edition in 1904, Volume 2, 119–122. Available at `http://www.hti.umich.edu/cgi/t/text/text-idx?c=umhistmath;idno=ACA7836`.
7. K. Böröczky: Packing of spheres in spaces of constant curvature. *Acta Math. Acad. Sci. Hungar.* 32 (1978) 243–261.
8. K. Böröczky Jr.: Finite packing and covering, *Cambridge University Press*, Cambridge, 2004.
9. P. Brass, W. Moser, and J. Pach: Research Problems in Discrete Geometry, *Springer-Verlag*, New York, 2005.
10. L. G. Casado, I. García, and Ya. D. Sergeyev: Interval Branch and Bound algorithm for finding the first-zero-crossing-point in one-dimensional functions. *Reliable Computing* 6 (2000) 179–191.
11. L. G. Casado, I. García, P. G. Szabó, and T. Csendes: Packing Equal Circles in a Square II. — New results for up to 100 circles using the TAMSASS-PECS stochastic algorithm. In: *Optimization Theory: Recent Developments from Mátraháza*, Kluwer, Dordrecht, 207–224, 2001.

12. J. H. Conway and N. J. A. Sloane: *Sphere Packings, Lattices and Groups*, Springer-Verlag, New York, 1998.
13. D. Cox, J. Little, and D. O'Shea: *Ideals, Varieties, and Algorithms.* An Introduction to Computational Algebraic Geometry and Commutative Algebra, Springer-Verlag, New York, 1997.
14. H. T. Croft, K. J. Falconer, and R. K. Guy: *Unsolved Problems in Geometry*, Springer-Verlag, New York, 108–110, 1991.
15. A. E. Csallner, T. Csendes, and M. Cs. Markót: Multisection in Interval Methods for Global Optimization I. Theoretical Results. *J. Global Optimization* 16 (2000) 371–392.
16. T. Csendes: Nonlinear parameter estimation by global optimization – efficiency and reliability. *Acta Cybernetica* 8 (1988) 361–370.
17. T. Csendes and D. Ratz: Subdivision direction selection in interval methods for global optimization. *SIAM J. Numerical Analysis* 34 (1997) 922–938. Available at `http://epubs.siam.org/fulltext/SINUM/volume-34/28152.pdf`.
18. DingZhu Du and P. M. Pardalos: *Minimax and Applications*, Kluwer, Dordrecht, 1995.
19. Z. Drezner, E. Erkut: Solving the Continuous p-dispersion Problem Using Non-linear Programming. *J. Operat. Research Soc.* 46 (1995) 516–520.
20. G. Dueck and T. Scheuer: Threshold accepting: a general purpose optimization algorithm appearing superior to simulated annealing. *J. Comput. Phys.* 90 (1990) 161–175.
21. L. Fejes Tóth: Über einen geometrischen Satz. In German. *Mathematische Zeitschrift* 46 (1940) 78–83.
22. L. Fejes Tóth: Punktverteilungen in einem Quadrat. In German. *Studia Sci. Math. Hungar.* 6 (1971) 439–442.
23. L. Fejes Tóth: *Lagerungen in der Ebene, auf der Kugel und im Raum.* In German. Springer-Verlag, Berlin, 1972.
24. G. Fejes Tóth: *Handbook of Discrete and Computational Geometry.* CRC Press, Boca Raton, 1997.
25. A. Florian: Verteilung von Punkten in einem Quadrat. In German. *Österreich. Akad. Wiss. Math. Natur. Kl. Sitzungsber. II* 198 (1989) 27–44.
26. F. Fodor: The densest packing of 19 congruent circles in a circle. *Geometriae Dedicata* 74 (1999) 139–145.
27. F. Fodor: The densest packing of 12 congruent circles in a circle. *Beiträge zur Algebra und Geometrie, Contributions to Algebra and Geometry* 41 (2000) 401–409.
28. F. Fodor: The densest packing of 13 congruent circles in a circle. *Beiträge zur Algebra und Geometrie, Contributions to Algebra and Geometry* 44:2 (2003) 431–440.
29. J. H. Folkman and R. L. Graham: A packing inequality for compact convex subsets of the plane. *Canadian Mathematical Bulletin* 12 (1969) 745–752.
30. H. Fukagawa and D. Pedoe: *Japanese temple geometry problems = Sangaku.* Charles Babbage Research Centre, Winnipeg, Canada, 1989.
31. Zs. Gáspár and T. Tarnai: Upper bound of density for packing of equal circles in special domains in the plane, *Periodica Polytechnica Ser. Civ. Eng.* 44:1 (2000) 13–32. Available at `http://www.pp.bme.hu/ci/2000_1/pdf/ci2000_1_02.pdf`.
32. M. Goldberg: The packing of equal circles in a square. *Mathematics Magazine* 43 (1970) 24–30.
33. M. Goldberg: Packing of 14, 16, 17 and 20 circles in a circle. *Mathematics Magazine* 44 (1971) 134–139.
34. J. E. Goodman and J. O'Rourke (eds.): *Handbook of Discrete and Computational Geometry*, CRC Press, New York, 1997.

35. R. L. Graham and B. D. Lubachevsky: Dense packings of equal disks in an equilateral triangle from 22 to 34 and beyond, *The Electronic J. of Combinatorics* 2, 1995. Available at `http://www.emis.de/journals/EJC/Volume_2/PDFFiles/v2i1a1.pdf`.
36. R. L. Graham and B. D. Lubachevsky: Repeated Patterns of Dense Packings of Equal Circles in a Square, *The Electronic J. of Combinatorics* 3 (1996) #R16, 211–227. Available at `http://www.combinatorics.org/Volume_3/PDFFiles/v3i1r16.pdf`.
37. R. L. Graham, B. D. Lubachevsky, K. J. Nurmela, and P.R.J. Östergård: Dense packings of congruent circles in a circle. *Discrete Mathematics* 181 (1998) 139–154.
38. M. Grannell: An even better packing of ten equal circles in a square. *Manuscript*, 1990.
39. C. de Groot, R. Peikert, and D. Würtz: The Optimal Packing of Ten Equal Circles in a Square. *IPS Research Report, Eidgenössiche Technische Hochschule, Zürich*, No. 90–12, August, 1990.
40. B. Grünbaum: An improved packing of ten circles in a square. *Manuscript*, 1990.
41. H. Hadwiger: Über extremale Punktverteilungen in ebenen Gebieten. *Math. Zeitschrift* 49 (1944) 370–373.
42. T. C. Hales: Cannonballs and Honeycombs. *Notices Amer. Math. Soc.* 47 (2000) 440–449.
43. R. Hammer, M. Hocks, U. Kulisch, and D. Ratz: *Numerical Toolbox for Verified Computing I.*, Springer-Verlag, Berlin, 1993.
44. E. Hansen: *Global Optimization Using Interval Analysis*, Marcel Dekker, New York, 1992.
45. T. Harayama: *Optimal Packings of 8, 9, and 10 Equal Circles in an Isosceles Right Triangle*, Master's thesis, School of Information Science, Japan Advanced Institute of Science and Technology, Japan, 2000.
46. P. van Hentenryck, D. McAllester, and D. Kapur: Solving Polynomial Systems Using a Branch and Prune Approach, *SIAM J. on Numerical Analysis* 34 (1997), 797–827.
47. R. Horst, P. M. Pardalos, and N. V. Thoai: *Introduction to Global Optimization*, Kluwer, Dordrecht, 1995.
48. R. Horst and N. V. Thoai: D.C. Programming: Overview. *J. of Optimization Theory and Applications* 103 (1999) 1–43.
49. M. Hujter: Some Numerical Problems in Discrete Geometry. *Computers and Mathematics with Applications* 38 (1999) 175–178.
50. D. C. Karnop: Random search techniques for optimization problems. *Automatica* 1 (1963) 111–121.
51. R. B. Kearfott: *Rigorous Global Search: Continuous Problems.* Kluwer, Dordrecht, 1996.
52. R. B. Kearfott: Test Results for an Interval Branch and Bound Algorithm for Equality-Constrained Optimization. In: *Computational Methods and Applications*, Kluwer, Dordrecht, 1996, 181–200.
53. K. Kirchner and G. Wengerodt: Die dichteste Packung von 36 Kreisen in einem Quadrat. *Beiträge zur Algebra und Geometrie, Contributions to Algebra and Geometry.* In German. 25 (1987) 147–159.
54. O. Knüppel: *PROFIL – Programmer's Runtime Optimized Fast Interval Library.* Bericht 93.4, Technische Universität Hamburg-Harburg, 1993.
55. O. Knüppel: *A Multiple Precision Arithmetic for PROFIL.* Bericht 93.6, Technische Universität Hamburg-Harburg, 1993.
56. O. Knüppel: *PROFIL/BIAS V 2.0*, Bericht 99.1, Technische Universität Hamburg-Harburg, 1999.
57. S. Kravitz: Packing cylinders into cylindrical containers. *Mathematics Magazine* 40 (1967) 65–71.

58. J. L. Lagrange: Recherches d'arithmétique. *Nouv. Mem. Acad. Roy. Sc. Belle Lettres*, Berlin, in Oeuvres III (1773) 693–758.
59. H. Lob and H. W. Richmond: On the solutions of Malfatti's problem for a triangle. *Proc. London Math. Soc.* 30 (1930) 287–304.
60. M. Locatelli and U. Raber: A Deterministic Global Optimization Approach for Solving the Problem of Packing Equal Circles in a Square. *International Workshop on Global Optimization (GO.99)*, Firenze, Italy, 1999.
61. M. Locatelli and U. Raber: Packing equal circles in a square: a deterministic global optimization approach. *Discrete Applied Mathematics* 122 (2002) 139–166.
62. B. D. Lubachevsky: How to simulate billiards and similar systems. *J. Computational Physics* 94 (1991) 255–283.
63. B. D. Lubachevsky and R. L. Graham: Curved Hexagonal Packings of Equal Disks in a Circle. *Discrete & Computational Geometry* 18 (1997) 179–194.
64. B. D. Lubachevsky, R. L. Graham, and F. H. Stillinger: Patterns and Structures in disk packings. *Periodica Mathematica Hungarica* 34 (1997) 123–142.
65. B. D. Lubachevsky and F. H. Stillinger: Geometric properties of random disk packings. *J. Statistical Physics* 60 (1990) 561–583.
66. C. D. Maranas, C. A. Floudas, and P. M. Pardalos: New results in the packing of equal circles in a square. *Discrete Mathematics* 142 (1995) 287–293.
67. M. Cs. Markót: An Interval Method to Validate Optimal Solutions of the "Packing Circles in a Unit Square" Problems. *Central European J. of Operational Research*, 8 (2000) 63–78. Available at `http://www.sztaki.hu/ markot/packcirc.htm`
68. M. Cs. Markót: Optimal Packing of 28 Equal Circles in a Unit Square – the First Reliable Solution. *Numerical Algorithms* (37) 2004, 253–261.
69. M. Cs. Markót: Interval Methods for Verifying Structural Optimality of Circle Packing Configurations in the Unit Square. To appear in J. Computational and Applied Mathematics.
70. M. Cs. Markót and T. Csendes: A New Verified Optimization Technique for the "Packing Circles in a Unit Square" Problems. *SIAM J. Optimization* 16 (2005), 193–219.
71. M. Cs. Markót, T. Csendes, and A. E. Csallner: Multisection in Interval Methods for Global Optimization II, Numerical Tests, *J. Global Optimization*, 16 (2000) 219–228.
72. J. Matyas: Random optimization. *Automatization and Remote Control*, 26 (1965) 244–251.
73. J. R. McDonnell and D. Waagen: Evolving recurrent perceptrons for time-series modeling, *IEEE Trans. on Neural Networks* 5 (1994) 24–38.
74. H. Melissen: Densest packings of congruent circles in an equilateral triangle. *American Mathematical Monthly* 100 (1993), 916–925.
75. H. Melissen: Densest Packing of Six Equal Circles in a Square, *Elemente der Mathematik* 49 (1994) 27–31.
76. J. B. M. Melissen: Densest packing of eleven congruent circles in a circle. *Geometriae Dedicata* 50 (1994) 15–25.
77. J. B. M. Melissen: Optimal packings of eleven equal circles in an equilateral triangle, *Acta Mathematica Hungarica* 65 (1994) 389–393.
78. J. B. M. Melissen and P. C. Schuur: Packing 16, 17 or 18 circles in an equilateral triangle. *Discrete Mathematics* 145 (1995) 333–342.
79. R. Milano: Configurations optimales de disques dans un polygone régulier, *Mémoire de Licence*, Université Libre de Bruxelles, 1987.
80. M. Mollard and C. Payan: Some progress in the packing of equal circles in a square. *Discrete Mathematics* 84 (1990) 303–307.

81. R. E. Moore: *Interval Analysis*. Prentice-Hall, Englewood Cliffs, 1966.

82. L. Moser: Problem 24 (corrected). *Canadian Mathematical Bulletin* 8 (1960) 78.

83. A. Neumaier: *Interval Methods for Systems of Equations*, Encyclopedia of Mathematics and its Applications 37, Cambridge Univ. Press, Cambridge, 1990.

84. A. Neumaier: *Introduction to Numerical Analysis*. Cambridge Univ. Press, Cambridge, 2001.

85. K. J. Nurmela: Constructing combinatorial designs by local search. Series A: Research Reports 27, Digital Systems Laboratory, Helsinki University of Technology, 1993.

86. K. J. Nurmela and P. R. J. Östergård: Packing up to 50 Equal Circles in a Square. *Discrete & Computational Geometry* 18 (1997) 111–120.

87. K. J. Nurmela and P. R. J. Östergård: More Optimal Packings of Equal Circles in a Square. *Discrete & Computational Geometry* 22 (1999) 439–457.

88. K. J. Nurmela and P. R. J. Östergård: Optimal packings of equal circles in a square. In Y. Alavi, D. R. Lick, and A. Schwenk (eds.): *Combinatorics, Graph Theory, and Algorithms (Proc. 8th Quadrennial International Conference on Graph Theory, Combinatorics, Algorithms, and Applications)*, 1999, 671–680.

89. K. J. Nurmela, P. R. J. Östergård, and R. aus dem Spring: Asymptotic Behavior of Optimal Circle Packings in a Square. *Canadian Mathematical Bulletin* 42 (1999) 380–385.

90. N. Oler: An inequality in the geometry of numbers. *Acta Math.* 105 (1961) 19–48.

91. N. Oler: A finite packing problem. *Canadian Mathematical Bulletin* 4 (1961) 153–155.

92. Ch. Payan: Empilement de cercles égaux dans un triangle équilatéral. Á propos d'une conjecture d'Erdős-Oler. In French. *Discrete Mathematics* 165/166 (1997) 555–565.

93. R. Peikert: Dichteste Packungen von gleichen Kreisen in einem Quadrat. In German. *Elemente der Mathematik* 49 (1994), 16–26.

94. R. Peikert, D. Würtz, M. Monagan, and C. de Groot: Packing Circles in a Square: A Review and New Results. In: P. Kall (ed.): *System Modelling and Optimization, Lecture Notes in Control and Information Sciences*, Springer-Verlag, Berlin, 180 (1992) 45–54.

95. J. Petris and N. Hungerbüler: *Manuscript*, 1990.

96. U. Pirl: Der Mindestabstand von n in der Einheitskreisscheibe gelegenen Punkten. In German. *Math. Nachr.* 40 (1969) 111–124.

97. U. Raber: *Nonconvex All-Quadratic Global Optimization Problems: Solution Methods, Application and Related Topics*, PhD thesis, University of Trier, 1999.

98. S. S. Rao: *Optimization Theory and Applications*. John Wiley and Sons, New York, 1978.

99. H. Ratschek and J. Rokne: *Computer Methods for the Range of Functions*. Ellis Horwood, Chichester, 1984.

100. H. Ratschek and J. Rokne: *New Computer Methods for Global Optimization*. Ellis Horwood, Chichester, 1988.

101. G. E. Reis: Dense packings of equal circles within a circle. *Mathematics Magazine* 48 (1975) 33–37.

102. M. Ruda: Körelhelyezések téglalapokon. In Hungarian. *MTA III. Osztály Közleményei* 19 (1969) 73–87.

103. J. Schaer: The densest packing of nine circles in a square. *Canadian Mathematical Bulletin* 8 (1965) 273–277.

104. J. Schaer: On the packing of ten equal circles in a square. *Mathematics Magazine* 44 (1971) 139–140.

105. J. Schaer and A. Meir: On a geometric extremum problem. *Canadian Mathematical Bulletin* 8 (1965) 21–27.

106. A. Schürmann: On Extremal Finite Packings. *Discrete & Computational Geometry* 28 (2002) 389–403.

107. K. Schlüter: Kreispackung in Quadraten. In German. *Elemente der Mathematik* 34 (1979) 12–14.
108. B. L. Schwartz: Separating points in a square. *J. of Recreational Mathematics* 3 (1970) 195–204.
109. F. J. Solis and J. B. Wets: Minimization by random search techniques. *Mathematics of Operations Research* 6 (1981) 19–50.
110. E. Specht's packing website. `http://www.packomania.com`.
111. R. J. Speedy: Random jammed packings of hard discs and spheres. *J. Phys.: Condens. Matter* 10 (1998) 4185–4194.
112. Gy. Staar: *A megélt matematika.* In Hungarian. Gondolat, Budapest, Hungary, 1990.
113. P. G. Szabó: Optimális körelhelyezések a négyzetben. In Hungarian. *Polygon*, X (2000) 48–64, 2000. Available at `http://www.inf.u-szeged.hu/~pszabo/Pub/polygon.ps.gz`.
114. P. G. Szabó: Some new structures for the "equal circles packing in a square" problem. *Central European J. of Operations Research* 8 (2000) 79–91.
115. P. G. Szabó: Sangaku, matematikai fatáblák japán templomokban. In Hungarian. *KöMaL* 7 (2001) 386–388. Available at `http://www.komal.hu/cikkek/sangaku/sangaku.h.shtml`.
116. P. G. Szabó: Optimal substructures in optimal and approximate circle packings. *Beiträge zur Algebra und Geometrie, Contributions to Algebra and Geometry* 46 (2005) 103–118.
117. P. G. Szabó and T. Csendes: Lázár Dezső és a kongruens körök négyzetben való legsűrűbb pakolásának problémája. In Hungarian. *Magyar Tudomány* 8 (2001) 984–985. Available at `http://www.matud.iif.hu/01aug/szabopg.html`.
118. P. G. Szabó, T. Csendes, L. G. Casado, and I. García: Packing Equal Circles in a Square I. – Problem Setting and Bounds for Optimal Solutions. In: *Optimization Theory: Recent Developments from Mátraháza*, Kluwer, Dordrecht, 191–206, 2001.
119. P. G. Szabó, M. Cs. Markót, and T. Csendes: Global Optimization in Geometry – Circle Packing into the Square, *Essays and Surveys in Global Optimization (Gerad 25th Anniversary Series)*, Ed. by C. Audet, P. Hansen, and G. Savard, Springer, 233–266, 2005.
120. P. G. Szabó and E. Specht: Packing up to 200 equal circles in a square. Submitted for publication. Available at `http://www.inf.u-szeged.hu/~pszabo/Pub/pack200.pdf`.
121. T. Tarnai: Packing of equal circles in a circle. *Structural Morphology: Toward the New Millenium*, The University of Nottingham, Nottingham, UK, 217–224, 1997.
122. T. Tarnai and Zs. Gáspár: Packing of equal circles in a square. *Acta Technica Acad. Sci. Hung.* 107 (1995–96) 123–135.
123. A. Thue: Om nogle geometrisk-taltheoretiske theoremer. In Norvegian. *Forhdl. Skand. Naturforskeres* 14 (1892) 352–355.
124. A. Thue: Über die dichteste Zusammenstellung von kongruenten Kreisen in einer Ebene. In German. *Norske Vidensk. Selsk. Skr.* 1 (1910) 3–9.
125. B. Tóth and T. Vinkó: Egy hatékony számítógépes eszköz matematikai problémák megoldására. In Hungarian. *Polygon* XI/1 (2002) 19–42. Available at `http://www.inf.u-szeged.hu/~boglarka/gspoly.ps.gz`.
126. G. Valette: A better packing of ten circles in a square. *Discrete Mathematics* 76 (1989) 57–59.
127. G. Wengerodt: Die dichteste Packung von 16 Kreisen in einem Quadrat. In German. *Beiträge zur Algebra und Geometrie, Contributions to Algebra and Geometry* 16 (1983) 173–190.

128. G. Wengerodt: Die dichteste Packung von 14 Kreisen in einem Quadrat. In German. *Beiträge zur Algebra und Geometrie, Contributions to Algebra and Geometry* 25 (1987) 25–46.
129. G. Wengerodt: Die dichteste Packung von 25 Kreisen in einem Quadrat. In German. *Ann. Univ. Sci. Budapest Eötvös Sect. Math.* 30 (1987) 3–15.
130. D. Würtz, M. Monagan, and R. Peikert: The history of packing circles in a square. *Maple Technical Newsletter* 0 (1994) 35–42.
131. V. A. Zalgaller and G. A. Los': The solution of the Malfatti-problem. In Russian. *Ukr. Geom. Sb.* 35 (1992) 14–33. English translation: *J. Math. Sci.* 72 (1994) 3163–3177.

Related websites

Program codes

Demonstration program for circle packing in Java:
`http://www.inf.u-szeged.hu/~csendes/pack.html`

Source code and the control scripts of the sophisticated discussed interval algorithm:
`http://www.sztaki.hu/~markot/packcirc.htm`

Source code, Makefile and the 'nice' library of the PDS algorithm:
`ftp://hydra.nat.uni-magdeburg.de/pub/packing/csq/csq.c`
`ftp://hydra.nat.uni-magdeburg.de/pub/packing/csq/Makefile`
`ftp://hydra.nat.uni-magdeburg.de/pub/nice/nice-2.0.tgz`

Packing websites

Packing website of E. Specht:
`http://www.packomania.com`

Packing website of P. G. Szabó:
`http://www.inf.u-szeged.hu/~pszabo/Packing_circles.html`

Papers

Papers by L. G. Casado:
`http://www.ace.ual.es/~leo/curr.html`

Papers by T. Csendes:
`http://www.inf.u-szeged.hu/~csendes/publ.html`

Papers by I. García:
`http://www.ace.ual.es/~inma/`

Papers by M. Cs. Markót:
`http://www.sztaki.hu/~markot/`

Papers by P. G. Szabó:
`http://www.inf.u-szeged.hu/~pszabo/publikacios_jegyzek.html`

Some more web pages

Erich's Packing Center:
`http://www.stetson.edu/~efriedma/packing.html`

M. Locatelli's home page:
`http://www.di.unito.it/~locatell/curr.html`

Circle packings by R. Peikert:
`http://www.cg.inf.ethz.ch/~peikert/personal/CirclePackings/`

H. Pfoertner, The Densest Packing of Equal Spheres in a Cube:
`http://www.randomwalk.de/sphere/incube/`

Zs. Ruttkay, Growing water lilies:
`http://wwwhome.cs.utwente.nl/~zsofi/sangaku/SanSim.html`

T. Tarnai's home page:
`http://www.epito.bme.hu/me/htdocs/dolgozok/`
`/dolgozo_reszlet.php?felhasznalonev=ttarnai`

T. C. Woo's works:
`http://depts.washington.edu/ie/people/woo.htm`
`http://depts.washington.edu/ie/people/faculty/`
`twoo/NickelToolbox.exe`

Al Zimmermann's Programming Contests:
`http://www.recmath.org/contest/`

List of Figures

List of Tables

Index

MIX
Papier aus verantwortungsvollen Quellen
Paper from responsible sources
FSC® C105338
FSC
www.fsc.org

If you have any concerns about our products,
you can contact us on
ProductSafety@springernature.com

In case Publisher is established outside the EU,
the EU authorized representative is:
Springer Nature Customer Service Center GmbH
Europaplatz 3, 69115 Heidelberg, Germany

Printed by Libri Plureos GmbH
in Hamburg, Germany